Physiological Effects Of Air Pollution

Volume V in MSS' series on Air Pollution

Papers by
Murray B. Gardner, David V. Bates, P. J. Lawther et al.

MSS Information Corporation
655 Madison Avenue, New York, N.Y. 10021

Library of Congress Cataloging in Publication Data

Gardner, Murray B   comp.
    Physiological effects of air pollution.

    CONTENTS: Gardner, M. B. and others. Pulmonary
changes in 7000 mice following prolonged exposure to
ambient and filtered Los Angeles air.--Bates, D. V.
Air pollutants and the human lung.--Lawther, P. J. and
others. Air pollution and exacerbations of bronchitis.
[etc.]
    1. Air--Pollution--Physiological effect--Addresses,
essays, lectures. I. Bates, David V., joint comp.
II. Lawther, P. J., joint comp. III. Title.
[DNLM: 1. Air pollution--Collected works. WA754
G227p 1973]

QP82.2.A3G37   616.2   73-10184
ISBN 0-8422-7137-6

TABLE OF CONTENTS

CREDITS AND ACKNOWLEDGEMENTS

Ayres, Stephen M.; and Meta E. Buehler, "The Effects of Urban Air Pollution on Health," *Clinical Pharmacology and Therapeutics*, 1970, 11:337-371.

Bates, David V., "Air Pollutants and the Human Lung," *American Review of Respiratory Disease*, 1972, 105:1-13.

Dahlhamn, Tore, "*In vivo* and *In vitro* Ciliotoxic Effects of Tobacco Smoke," *Archives of Environmental Health*, 1970, 21:633-634.

Emik, L. Otis; Roger L. Plata; Kirby I. Campbell; and George L. Clarke, "Biological Effects of Urban Air Pollution," *Archives of Environmental Health*, 1971, 23:335-342.

Ferris, Benjamin G., "Tests to Assess Effects of Low Levels of Air Pollutants on Human Health," *Archives of Environmental Health*, 1970, 21:553-558.

Freeman, Aaron E.; Paul J. Price; Robert J. Bryan; Robert J. Gordon; Raymond V. Gilden; Gary J. Kelloff; and Robert J. Huebner, "Transformation of Rat and Hamster Embryo Cells by Extracts of City Smog," *Proceedings of the National Academy of Sciences*, 1971, 68:445-449.

Gardner, Murray B.; Clayton G. Loosli; Bernard Hanes; William Blackmore; and Dixie Teebken, "Pulmonary Changes in 7,000 Mice Following Prolonged Exposure to Ambient and Filtered Los Angeles Air," *Archives of Environmental Health*, 1970, 20:310-317.

Heimbach, James A., Jr., "The Correlation of Polluted Air with Tree Growth and Lung Disease in Humans," *Computers in Biology and Medicine*, 1971, 1:243-253.

Lave, Lester B.; and Eugene P. Seskin, "Air Pollution and Human Health," *Science*, 1970, 169:723-733.

Lawther, P.J.; R.E. Waller; and Maureen Henderson, "Air Pollution and Exacerbations of Bronchitis," *Thorax*, 1970, 25:525-539.

Morris, S.C.; and M.A. Shapiro, "Statistical Note on Association of Air Pollution and Liver Cirrhosis," *Archives of Environmental Health*, 1972, 24:290-292.

PREFACE

In the United States air pollution reaches toxic thresholds in almost all urban areas. Its cost in terms of life and health is extreme. According to various governmental studies, the annual cost of damage resulting from polluted air is over $13 billion. If, for example, pollution could be cut by 50 percent, newborn babies would have an additional life expectancy of three to five years and deaths would be reduced by 4.5 percent. Air pollution is also a factor in the development of diseases ranging from respiratory disorders to cirrhosis of the liver, and polluted air intensifies allergic reactions. Serious long-term research and conscientious political action are needed if these ominous trends are to be reversed.

Volume V in MSS' continuing series on air pollution presents a group of current papers on the physiological effects of pollution. Because pollution is closely linked to various lung diseases such as bronchitis, studies on the cellular and pulmonary changes due to exposure to urban air are included. The long term effects of low levels of air pollution on human health are also discussed in detail.

Physiological Effects of Air Pollution

Pulmonary Changes in 7,000 Mice Following Prolonged Exposure to Ambient and Filtered Los Angeles Air

Murray B. Gardner, MD; Clayton G Loosli, MD; Bernard Hanes, PhD; William Blackmore, DVM; and Dixie Teebken, MA, Los Angeles

This report summarizes the incidence and appearance of lung tumors and pneumonitis in over 7,000 mice, following prolonged exposure to ambient as compared with filtered Los Angeles air. Mice in the ambient air colonies showed no difference in histologic appearance and no increase in incidence of lung tumors in two lung tumor susceptible and one lung tumor resistant strains; on the contrary, more lung adenomas in A/J mice were noted in the filtered air group. Acute bronchopneumonia and interstitial pneumonitis were qualitatively similar but quantitatively significantly more common in ambient air C57 black mice.

The findings suggest that prolonged exposure to ambient Los Angeles air is associated in several strains of mice with an increased susceptibility to pulmonary infection but not to increased pulmonary neoplasia.

THIS PAPER reports the long-term effect of inhaling Los Angeles ambient air upon the incidence and appearance of lung tumors and pneumonitis in mice. Over 7,000 mice of three inbred strains differing in their genetic susceptibility to lung tumors have been studied over the last five years. Although a lung tumor promoting activity of Los Angeles urban air has not been demonstrated, there is statistical evidence that residence in ambient air is associated with an increased susceptibility to pulmonary infection.

Materials and Methods

The design of the exposure and filtered air rooms, the air monitoring instrumentation utilized, and the air quality measurements recorded have been presented in previous publications.[1,2] Cleansed air was furnished by filtering the ambient air through an activated charcoal and fine pore filter for the removal of particles over 0.3μ in diameter. Nitrogen dioxide, sulfur dioxide, ozone and other oxidants, and many unreacted hydrocarbons were effectively removed by these filters. On the other hand, carbon monoxide, nitric oxide, and short-chain hydrocarbons (eg, methane) have been practically unaffected by the filter units. Slightly greater ranges of temperature and humidity were recorded in the ambient air rooms.

Three inbred strains of mice were used: Lung tumor susceptible A strain and A/J strain, between 1963 and 1965, and lung tumor resistant C57 black strain, from 1965 to 1967. Colonies of mice were divided into equal groups of ambient air and filtered air animals of equivalent age and sex, with litter mates identified. Mice were kept in wire mesh cages in groups of ten, segregated by sex. All animals were about 6 weeks of age at the beginning of the exposure.

Most of the strain A and A/J mice were periodically killed for microscopic study. Following spontaneous death, microscopic examination was also done on all mice that were not subjected to undue postmortem autolysis or

Submitted for publication June 9, 1969; accepted June 19.

From the departments of pathology (Drs. Gardner, Loosli, and Hanes) and vivaria (Dr. Blackmore and Miss Teebken), University of Southern California School of Medicine, Los Angeles.

Reprint requests to 2025 Zonal Ave, Los Angeles 90033 (Dr. Gardner).

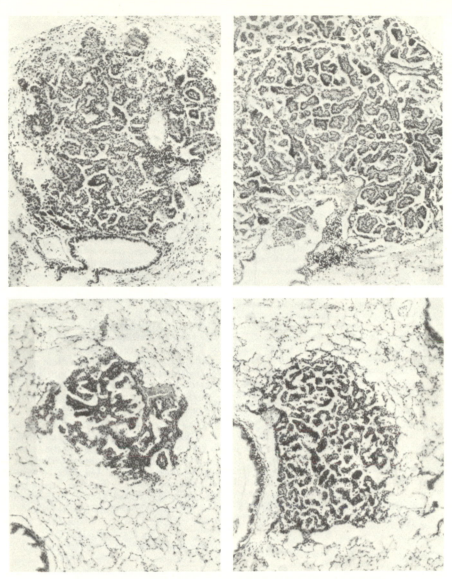

Fig 1.—Pulmonary tumors in A/J and C57 black strain mice. Tumors observed are adenomatous proliferations of alveolar wall cells. There is no histologic difference between tumors arising in ambient vs filtered air mice. There is tendency for tumors in A/J strain mice to be slightly larger than those in C57 black mice. **Top left,** Ambient air, A/J male mouse, age 17 months (hematoxylin and eosin, slightly reduced from X 125). **Top right,** Filtered air, A/J female mouse, age 15 months (hematoxylin and eosin, slightly reduced from X 125). **Bottom left,** Ambient air, C57 male mouse, age 20 months (hematoxylin and eosin, slightly reduced from X 125). **Bottom right,** Filtered air, C57 male mouse, age 21 months (hematoxylin and eosin, slightly reduced from X 125).

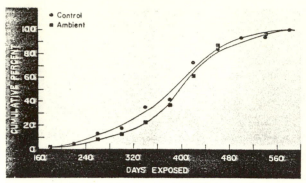

Fig 2.—Cumulative percent of A/J killed mice with lung tumors vs days exposed.

cannibalization. Aliquots of approximately 60 A and A/J strain mice from each atmosphere were randomly chosen for killing with ether at nearly monthly intervals, between 6 and 19 months of age. The entire large C57 black colony was examined subsequent to spontaneous (attritional) death. The lungs of all mice, killed and attritional, were inflated in vivo with a tracheal catheter using 10% buffered formaldehyde solution. After fixation in 10% buffered formaldehyde solution and embedding in paraffin, five-micron sections were cut and stained with hematoxylin and eosin. Several transverse sections extending from hilus to lateral lung margin were taken from each pulmonary lobe and aligned on the slide in a uniform manner using an agar technique.[3] In this fashion, 8 to 15 lung sections from each mouse were appropriately oriented so as to insure histologic confirmation of all pulmonary nodules and consolidations. Hilar lymph nodes were sometimes included, but no special attempt was made to examine all regional lymph nodes. Other organs from killed A strain and A/J mice and from C57 colonies were not sectioned. Death weights were recorded and the presence of pulmonary tumors or pneumonitis and other pertinent histologic features were recorded on cards for computer retrieval and analysis.

Results

Lung Tumors.—*Description.*—The lung tumors in each strain of mouse consisted of a uniform gross and microscopic appearance, as described in many prior studies.[4,5] They were usually white, discrete round nodules, peripherally located, 1 mm or less in diameter. Many were too small to be detected by the naked eye. There was no predilection for side, lobe, or sex. The microscopic appearance was that of a nonencapsulated nodular adenomatous proliferation of uniform alveolar cells with infrequent mitoses. Papillary fronds and small cysts were frequent components of the larger tumors. Expansile growth of the tumors caused some compression of the adjacent lung but there was little or no host inflammatory response about the tumor margins.

No metastases in hilar or mediastinal adenomas did occasionally extend into distal bronchioles (Fig 1, *top right*). Tumors bore no apparent topographic relationship to foci of pulmonary inflammation. Histopathologic features of incipient or overt squamous cell carcinoma were not seen.

The individual tumors in A/J mice were often slightly larger and slightly more cytologically pleomorphic than seen in A strain or C57 black mice. However, within each strain, tumors were of equal size and equivalent histologic appearance in both ambient and filtered air groups (Fig 1).

Incidence.—The overall incidence of lung tumors in the three strains of mice is shown in the Table. The genetically tumor-susceptible A and A/J strain mice showed no increase in total adenoma-bearing mice in the ambient air group. Quite to the contrary, study of a large number of killed A/J mice between 6 and 19 months of age showed that the filtered air group experienced a statistically significant increased overall lung adenoma incidence compared to the ambient air group (32% vs 20%, $P = < 0.001$). In the A/J mouse study there was no difference in tumor incidence observed at the time of initial killing, age 6 to 7 months. Thereafter, the cumulative frequency of adenoma-bearing mice increased with age in both groups but to a slightly greater degree in the filtered air group (Fig 2).

In the smaller sized attritional group of A/J mice, by contrast, single and multiple tumor incidences were both greater in the

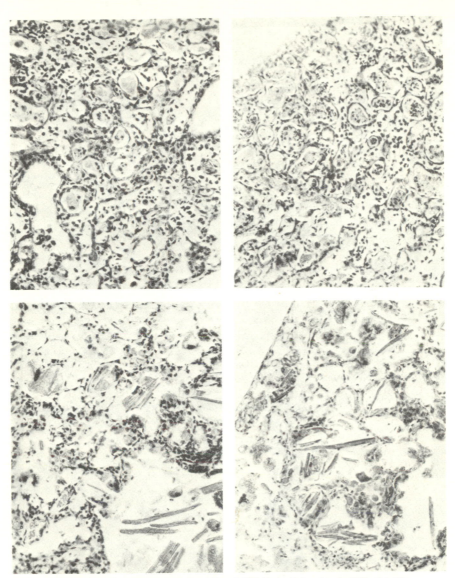

Fig 3.—Interstitial pneumonitis in A/J and C57 black strain mice. In C57 mice this desquamative inflammatory process is especially characterized by deposition of small and large needle-like eosinophilic crystals. There is, however, no difference within respective mouse strains in histologic appearance of interstitial pneumonitis between ambient and filtered air animals. **Top left,** Ambient air, A/J female mouse, age 12 months (hematoxylin and eosin, slightly reduced from ✕ 250). **Top right,** Filtered air, A/J male mouse, age 12 months (hematoxylin and eosin, slightly reduced from ✕ 250). **Bottom left,** Ambient air, C57 male mouse, age 28 months (hematoxylin and eosin, slightly reduced from ✕ 250). **Bottom right,** Filtered air, C57 male mouse, age 23 months (hematoxylin and eosin, slightly reduced from ✕ 250).

Mouse Exposure to Los Angeles Urban Air, 1963 to 1967

Dates Mice strain	1963-1964 A Strain, Killed		1963-1964 A Strain, Attritional		1963-1965 A/J Strain, Killed		1963-1965 A/J Strain, Attritional		1965-1967 C57 Strain, Attritional	
Exposure No. of mice (out of 7,248)	C* 516	A† 525	C 61	A 58	C 743	A 1057	C 252	A 208	C 1887	A 1941
Diagnosis Tumor-bearing mice	73 (14%)	77 (15%)	1	4	237 (32%)	213 (20%)	46 (18%)	62 (30%)	26 (1.3%)	34 (1.5%)
Mice bearing multiple tumors‡	0	0	0	0	13 (2%)	35 (3%)	10 (4%)	18 (9%)	2	0
Acute bronchopneumonia	0	1	0	4	17 (2%)	15 (1%)	59 (23%)	31 (15%)	178 (9%)	311 (16%)
Interstitial pneumonitis	202	218	18	15	211	468	22	37	Minimal	
	(39%)	(42%)	(30%)	(26%)	(29%)	(44%)	(9%)	(18%)	591 (31.3%)	544 (28%)
									Severe	
									162 (8.6%)	223 (11.4%)
Mean life-span (days)			Female				Female		Female	
			407 Male	411			455 Male	423	642 Male	655
			309	369			495	495	624	538
Mean death weight (gm)			Female				Female		Female	
			16 Male	16			16 Male	19	17 Male	17
			17	17			17	18	19	19

* Control, filtered air.
† Ambient air.
‡ Usually two tumors, never more than three per mouse.

ambient air group ($P = 0.01$ for single and $P = < 0.05$ for multiple adenomas). Multiple adenomas were usually two and never more than three per animal.

Killed A strain mice between 8 and 15 months of age showed an equal and surprisingly low overall incidence (15%) of pulmonary tumors, with an identical cumulative frequency curve in both groups and a complete lack of multiple adenoma-bearing mice.

The genetically tumor-resistant C57 colony showed the anticipated low incidence (1% to 2%) of tumors in the filtered air group, and no age-related increase in tumor incidence was noted in the ambient air group.

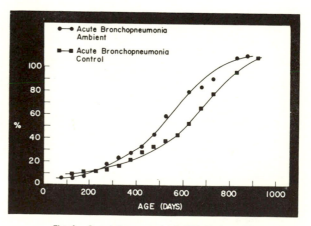

Fig 4.—Cumulative percent of C57 black attritional male mice with acute bronchopneumonia vs days exposed.

Pneumonitis.—*Description.*—Three main patterns of distinctive pneumonitis could be distinguished in each strain of mouse: (1) bronchopneumonia, (2) interstitial pneumonitis, and (3) embolic pneumonitis. Bronchopneumonia usually involved more than one lobe with an extensive polymorphonuclear exudate within bronchial passageways and adjacent alveoli. Interstitial pneumonitis usually presented as one or more microscopic sized foci of alveolar septal thickening with chronic inflammation, proliferation of alveolar cells, and histiocytic desquamation (Fig 3, *top left*, and *top right*). Occasionally this process was extensive, involving multiple foci, sometimes entire lobes. In the C57 black mice intra-alveolar desquamation of histiocytes, together with eosinophilic, needle-like microcrystals and macrocrystals[6] was particularly marked (Fig 3, *bottom left*, and *bottom right*). Bronchopneumonia was frequently superimposed upon interstitial pneumonitis. Embolic pneumonitis was distinguished by the presence of acute septic vasculitis surrounded by foci of acute inflammation.

Each of these patterns of pneumonitis was qualitatively similar and indistinguishable between ambient air and filtered air groups (Fig 3).

Conspicuously lacking in these mice was any evidence suggestive of chemical inflammation or pneumonitis, ie, vascular congestion, edema, or alteration in the mucosa or caliber of the proximal and distal airways. Anthracotic pigment deposition and emphysema were also absent.

Incidence (Table).—In the large C57 mouse colony there was a significant increase in incidence of acute bronchopneumonia, after age 8 months (Fig 4), from the ambient air group (16% vs 9%, $P = < 0.01$). From August to October 1964, an outbreak of acute bronchopneumonia caused the loss of a disproportionately large number of young A/J mice from one of the filtered air rooms.

Interstitial pneumonitis was significantly more common in ambient A/J mice, following either killing (44% vs 29%, $P = < 0.001$) or attritional death (18% vs 9%, $P = < 0.01$). There was also a tendency for severe degrees of interstitial pneumonitis to occur more frequently in ambient C57 mice (11.4% vs 8.6%, $P = < 0.05$), particularly in male mice. In the C57 mice both ambient

and filtered air groups had a 3% incidence of blood-borne embolic pneumonitis.

It is worthwhile noting that in the C57 mice squamous metaplasia and extensive bronchiolar-alveolar cell hyperplasia have been infrequent findings, always in association with severe interstitial pneumonitis. However, these features have been observed in 19 ambient air mice and only 1 filtered air mouse.

Life-Span and Death Weight.—There was no marked difference in life-span or death weight (Table) between ambient and filtered air groups within any of the mice colonies. In the C57 black and A strain colonies, female mice showed a longer mean life-span than did the male mice. The opposite trend was noted in the A/J colony in that male mice showed a longer mean life-span.

Comment

In our initial report[1] several years ago, we concluded that Los Angeles ambient air exposure might be promoting a slight increase in lung tumors in aging inbred mice. With the total study now completed and data thoroughly analyzed, it is clearly evident that prolonged exposure of laboratory mice to Los Angeles ambient air is *not* associated with an increase in lung tumors. The absence, in ambient air mice, of any increase in overall tumor incidence or of a shortened tumor latent period or of large and numerous tumors per mouse is evidence against any lung tumor promoting or accelerating influence.[5] Only in the relatively small group of A/J mice examined following attritional death was there an increased adenoma incidence in the ambient air group. This is probably explained by an increased death rate of young A/J mice in one filtered air room from a limited outbreak of acute bronchopneumonia, thus allowing proportionately more ambient air mice to live to older age with an increased opportunity to develop spontaneous tumors.

The 12% decrease (20% vs 32%) in overall tumor incidence of ambient as compared with filtered air mice observed in the much larger group of A/J mice examined following periodic killing is statistically quite significant ($P = < 0.001$). This finding, totally unanticipated, suggests that in this particular lung tumor susceptible strain, ambient air exposure may actually inhibit the appearance of lung tumors. Factors that have been associated with inhibition of urethan-induced and spontaneous lung tumors in mice include dietary restriction,[7] inhibition of deoxyribonucleic acid synthesis,[8,9] and survival following experimental infection with influenza virus.[10] Could it be that the significantly greater incidence of intercurrent interstitial pneumonitis, presumably of viral origin, in the ambient air A/J mice, has, by an unknown mechanism, similarly conferred upon this group some degree of resistance to lung tumor formation?

It was also contrary to our expectations to find such a low overall incidence (15%) of pulmonary tumors in A strain mice, traditionally the inbred strain of highest spontaneous lung tumor occurrence.[11] That A/J strain mice are apparently more prone to spontaneous lung tumors is shown by their greater overall tumor incidence, their tendency to exhibit slightly larger tumors, and the more frequent occurrence of two or three tumors in a single animal. It was hoped that by using a tumor-resistant strain, C57 black,[11] any increase in tumor incidence observed in ambient air as compared with filtered air mice would take on added significance, but this did not prove to be the case.

Our inability to detect any tumor-promoting effect upon mouse lung from inhalation of Los Angeles urban air is, perhaps, not too surprising when one considers that even peak pollutant levels in the ambient atmosphere are still several times less (about two to ten times) than those levels maintained in the *synthetic* atmospheres reported[12-14] to increase mouse lung tumor incidence. Other mitigating factors might be the relatively advanced age at initial exposure (6 weeks), the obligate nose breathing with resultant air "cleansing," the potential antagonistic interaction of smog components,[15] or the development of tolerance[16] to oxidant inhalants. Peacock[17] also failed to observe any increase in lung tumor incidence in C3H mice, a moderately susceptible strain, following lifelong exposure to ambient air of Glasgow, Scotland.

The mouse lung tumor is, by no means, a biologic counterpart of the human lung tumor. In contrast to the bronchogenic origin and malignant behavior of the usual human lung tumor, the mouse lung tumor is composed of an adenomatous proliferation of alveolar wall cells,[4] possessing the most indolent malignant potential, as supported by the total absence of metastases in this study. The spontaneous occurrence of such pulmonary adenomatous tumors in mouse lung is determined to some extent by the genetically controlled turnover rate of alveolar cells,[18] but it is independent of any known virus[5] or hormonal factors,[5] and this tumor is free of demonstrable tumor-specific antigens.[19] Equating this lesion with lung carcinogenicity in mice, or even more so, with lung carcinoma in man is, in our opinion, unwarranted. It should be noted that the absence of incipient or overt bronchogenic carcinoma in these mice is consistent with the tobacco smoke inhalation studies of Wynder and Hoffmann,[20] who also concluded that a better biologic assay than mouse lung is required for testing of inhalant carcinogenicity. The absence of an ambient air tumorigenic effect upon mouse lung is consistent with the failure to detect, in humans, any increased death rate from lung carcinoma that could be related to residence specifically in Los Angeles.[21]

Our evidence for an increased susceptibility to pulmonary infection in ambient air mice as compared with filtered air mice is entirely statistical, but it is of a sufficient degree of certainty to be biologically meaningful. An age-related increased susceptibility to bacterial pneumonitis is supported by the findings of more acute bronchopneumonia in the ambient C57 mice. Susceptibility to endemic viral pneumonitis is suggested by the increased incidence of "incidental" microscopic foci of interstitial pneumonitis in ambient air A/J mice and of severe interstitial pneumonitis in ambient air C57 mice. We have detected no morphologic explanation for this postulated increased susceptibility to infection. Transient or subtle changes, such as bronchospasm or edema or fine structural alterations,[22] could, of course, escape detection. Perhaps susceptibility is related to the somewhat greater swings in temperature and humidity recorded in the ambient air exposure rooms, or to the presence in the ambient rooms of more air-borne particulates and, possibly, microbial pathogens. There are several recent reports which show that oxidant inhalants in concentrations comparable to Los Angeles smog do indeed increase susceptibility to experimental,[23,24] and spontaneous[25] bacterial pneumonitis in rodents. An epidemiologic study[26] on the human population in Los Angeles has suggested a positive correlation between respiratory disease morbidity and air pollution levels.

The lack of detectable cumulative or chronic residua,[27] including tumors, in the lungs of mice breathing ambient air, suggests that, for this species and target organ, at any rate, natural atmospheric conditions in Los Angeles exert a relatively innocuous direct biological effect.

This investigation was supported in part by Public Health Service research contract PH-86-68-43 and by the Hastings Fund of University of Southern California School of Medicine and by the Council for Tobacco Research, United States.

References

1. Gardner, M.B.: Biological Effects of Urban Air Pollution: III. Lung Tumors in Mice, *Arch Environ Health* 12:305-313 (March) 1966.

2. Bryan, R.J.: Instrumentation for an Ambient Air Animal Exposure Project, *J Air Pollut Contr Assoc* 13:254-265 (June) 1963.

3. Shackel, R.I., and Jones, J.L.: An Embedding Medium for Permanent Sections of Exudative Material and Fragments of Material Removed for Biopsy, *Techn Bull Regist Med Techn* 29:155-156 (Sept) 1959.

4. Grady, H.G., and Stewart, H.L.: Histogenesis of Induced Pulmonary Tumors in Mice, *Amer J Path* 16:417-432 (July) 1940.

5. Shimkin, M.B.: Pulmonary Tumors in Experimental Animals, *Advances Cancer Res* 3:223-267, 1955.

6. Sherwin, R.P.; Gardner, M.B.; and Richters, V.: The Occurrence of Crystalline Deposits in the Lungs of C57 Mice, *Arch Environ Health* 15:589-598 (Nov) 1967.

7. Tannenbaum, A., and Silverstone, H.: Nutrition in Relation to Cancer, *Advances Cancer Res* 1:451-501, 1953.

8. Foley, W.A., and Cole, L.J.: Inhibition of Urethan Lung Tumor Induction in Mice by Total-Body X-Radiation, *Cancer Res* 23:1176-1180 (Sept) 1963.

9. Shimkin, M.B., et al: Relation of Thymidine Index to Pulmonary Tumor Response in Mice Receiving Urethane and Other Carcinogens, *Cancer Res* 29:994-998 (May) 1969.

10. Steiner, P.E., and Loosli, C.G.: The Effect of Human Influenza Virus (Type A) on the Incidence of Lung Tumors in Mice, *Cancer Res* 10:385-392 (June) 1950.

11. Heston, W.E.: Genetics of Cancer, *Advances Genet* 2:99-125, 1948.

12. Kotin, P., and Falk, H.L.: II. The Experimental Induction of Pulmonary Tumors in Strain-A Mice After Their Exposure to an Atmosphere of Ozonized Gasoline, *Cancer* 9:910-917 (Sept-Oct) 1956.

13. Kotin, P., and Falk, H.L.: III. The Experimental Induction of Pulmonary Tumors and Changes in the Respiratory Epithelium in C57BL Mice Following Their Exposure to an Atmosphere of Ozonized Gasoline, *Cancer* 11:473-481 (May-June) 1958.

14. Stokinger, H.E.: "Effects of Air Pollution on Animals," in Stern, A.C. (ed.): *Air Pollution*, New York: Academic Press, Inc., 1962, vol 1, pp 282-334.

15. Wagner, W.D.; Dobrogorski, O.J.; and Stokinger, H.E.: Antagonistic Action of Oil Mists on Air Pollutants: Effects on Oxidants, Ozone, and Nitrogen Dioxide, *Arch Environ Health* 2:523-534 (May) 1961.

16. Fairchild, E.J.: Tolerance Mechanisms: Determinants of Lung Responses to Injurious Agents, *Arch Environ Health* 14:111-125 (Jan) 1967.

17. Peacock, P.R.: "An Etiological Study of Lung Tumors in Mice," in Severi, L. (ed.): *The Morphological Precursors of Cancer*, Perugia, Italy: Division of Cancer Research, Perugia University Press, 1962, pp 605-610.

18. Simnett, J.D., and Heppleston, A.G.: Cell Renewal in the Mouse Lung: The Influence of Sex, Strain, and Age, *Lab Invest* 15:1793-1801 (Nov) 1966.

19. Law, L.W.: Studies of the Significance of Tumor Antigens in Induction and Repression of Neoplastic Disease, presidential address, *Cancer Res* 29:1-21 (Jan) 1969.

20. Wynder, E.L., and Hoffmann, D.: Experimental Tobacco Carcinogenesis, *Science* 162:862-871 (Nov 22) 1968.

21. Buell, P.; Dunn, J.E.; and Breslow, L.: Cancer of the Lung and Los Angeles-type Air Pollution, *Cancer* 20:2139-2147 (Dec) 1967.

22. Bils, R.F.: Ultrastructural Alterations of Alveolar Tissue of Mice: I. Due to Heavy Los Angeles Smog, *Arch Environ Health* 12:689-697 (June) 1966.

23. Ehrlich, R.: Effect of Nitrogen Dioxide on Resistance to Respiratory Infection, *Bact Rev* 30:604-614 (Sept) 1966.

24. Coffin, D.L., and Blommer, E.J.: Acute Toxicity of Irradiated Auto Exhaust: Its Indication by Enhancement of Mortality From Streptococcal Pneumonia, *Arch Environ Health* 15:36-38 (July) 1967.

25. Hueter, F.G., et al: Biological Effects of Atmospheres Contaminated by Auto Exhaust, *Arch Environ Health* 12:553-560 (May) 1966.

26. Sterling, T.D., et al: Urban Morbidity and Air Pollution, *Arch Environ Health* 13:158-170 (Aug) 1966.

27. Wayne, L.G., and Chambers, L.A.: Biological Effects of Urban Air Pollution: V. A Study of Effects of Los Angeles Atmosphere on Laboratory Rodents. *Arch Environ Health* 16:871-885 (June) 1968.

Air Pollutants and the Human Lung

The James Waring Memorial Lecture[1-3]

DAVID V. BATES

Introduction

Dr. Waring wrote extensively on different aspects of lung disease that dominated the field in his day. His papers reflect an inclination he had to look backward in time, as he did, for example, in his 1936 paper on the "Vicissitudes of Auscultation," and to look forward to new developments and new insight on old problems. In this spirit, and in his memory, I wish to look backward and forward at the problem of air pollution and the lung. The moment is propitious. I can only disclaim the accusation that I am taking advantage of popular concern about this problem by pointing out that in one way and another I have been involved with it for more than 15 years. I chose it primarily, however, because I believe that scientists working in the pulmonary field have recently acquired new knowledge about mechanisms of lung disease of importance in illuminating previous data and guiding

us in future research, and I wished to draw these to your attention.

I am aware of the hazards of an attempt to relate air pollution to lung disease. Epidemiologists are careful to exclude very much discussion on the possible pathophysiologic mechanisms from their papers, and in this they surely show good judgment. It is unfortunately true that some

[1] Department of Physiology, McGill University, Montreal, Canada, and Royal Victoria Hospital, Montreal, Canada.

[2] Delivered at University of Colorado, Denver, on March 29, 1971 and at Stanford University, Palo Alto, California, on April 1, 1971.

[3] The Waring Lectureship was established in 1965 by the Francis D. North Foundation in honor of James J. Waring, M.D., who was Professor of Medicine at the University of Colorado School of Medicine.

physicians have been inclined to discount the epidemiologic data because no one was able to put forward any very convincing evidence of pathogenetic mechanisms. Experiments relevant to an understanding of air pollutants are often missed by both of these groups of workers because such experimental data often seem to exist in a world of their own.

Some speakers on air pollution are fond of describing it as "a cause looking for a disease." This presumes that unless some distinctive clinical condition can be shown to be directly caused by air pollution, we can assume that it is without effect; I hope to do something to undermine such a simplistic view of disease.

We can recognize that any attempt to understand the relationship between air pollutants and the human lung must, if it is to be successful, depend on many areas of study, and it is for this reason that I have thought it worthwhile to attempt to see what kind of structure can be built up, making use of information now available to us. It is easy to bring forward reasons why one should not attempt any synthesis of existing information, but I hope to show that such a review may, at this point in time, throw into sharp focus some extrapolations we cannot make, some coincidences that may be important, some areas where a need for new data is particularly acute, as well as some simplistic ideas of lung disease that we would do well to abandon.

I do not feel competent to deal with any aspect of lung cancer relating to air pollution. This is such a specialized field that I will have to leave to others the task of discussing whether there is indeed an increased burden of lung cancer in urban communities, and if so, to what it is likely to be due. I am going to address myself only to chronic bronchitis and emphysema and to recent data on acute chest infections in school children and infants less than the age of two years, since this seems to me highly relevant to the problem.

Epidemiologic Data

The epidemiologic data to this point seem to provide evidence for five quite particular and specific statements. These are as follows:

(1) Where air pollution is sufficiently great, episodes of increased sulfur dioxide and particulate pollution result in an increased mortality among those with chronic lung disease (1) and some increase in morbidity (2). In exceptionally severe episodes such as occurred in London in 1952, there is also an increased mortality in younger age groups and even in infants. The concentrations of smoke and sulfur dioxide that occurred in London in 1952 are, one hopes, unlikely now to be repeated anywhere because control measures have substantially lowered sulfur dioxide and smoke concentrations in London (3) and in New York City (4).

(2) Concentrations of sulfur dioxide (SO_2) greater than approximately 0.08 ppm as an annual average, particularly when accompanied by particulate pollution of the order of more than 150 μg per m^3 as an annual average, are accompanied by an increased morbidity from chronic bronchitis in the adult city population (5). This seems to me to have been clearly established by the data of Holland and Reid (6) and more recent studies (4, 7).

(3) If proper account is taken of cigarette smoking, it is possible to show that the forced expiratory volume in one second (FEV_1) is lower in men of equivalent age with the same smoking history doing the same job in country towns of England where pollution is reasonably low compared to their opposite numbers in London (6). This decrement of FEV_1 is evident also in non-smokers, but the proportionate difference between the dweller in the low pollution city compared to the worker in the high pollution area is greatest in the heaviest cigarette smoking group of the population (7). We should note, however, that we do not know whether this "additive" effect of air pollution is exerted by producing a more severe degree of chronic bronchitis, or by more severe damage as a consequence of more frequent infection, or even by a more severe degree of concomitant emphysema in those living in the more polluted region.

(4) A careful study by Douglas and Waller (8) has shown beyond reasonable argument that there is a three-fold increase in morbidity from lower chest infections in in-

fants younger than two years in moderate and high pollution regions, compared to very low pollution regions. Their data show a significant increase in morbidity between regions of very low pollution (SO_2 less than 0.03 ppm) and areas with moderate pollution (annual concentrations of SO_2 of approximately 0.06 ppm). There is further, recent, very disquieting, evidence that a relationship exists between the concentration of oxides of nitrogen (at an annual average of more than 0.06 ppm) and morbidity from lower chest infection (9) in the same age group. This concentration of nitrogen oxides is exceeded in many cities in the United States at this point in time. We clearly have to account for this data if we are to reach any complete description of the effects of air pollutants on the lung during its period of growth and development. We do not know whether the attack rate for all these populations is similar, but the clinical manifestation is more severe in those living in more polluted regions; or whether the attack rate varies as a consequence of air pollution. Further, it is important to recognize that although the pediatrician serving an industrial population sees more cases of bronchiolitis in infants than his rural counterpart, there is no clinical distinction to be made between those occurring in cities and in rural areas except on a basis of frequency.

(5) Finally, there is evidence from Japan that symptoms and pulmonary function in school children in grossly polluted regions of that country (the reported concentrations of sulfur dioxide and particulate pollution of various kinds are extremely high in parts of Japan and probably approached in very few other places) are both measurably different as compared with those in children in low pollution regions (10). Recently Colley and Reid (11) found that in the United Kingdom there is apparently an economic factor built into the difference between school children in low and high polluted regions of that country, and it is only children in the lower economic group who show much significant difference in respiratory disease morbidity between the ages of 9 years and 11 years.

This necessarily brief review of a small number of selected references from what is a very large body of data must serve to orientate us toward the phenomena we have to explain. I am aware that some people have suggested that the "urban factor," which appears to be causing these phenomena, may have nothing to do with the direct effect of any pollutant on the lung, but is entirely accountable on the basis of the increased density of population, some potentiating effect of pollutants on bacterial growth in the air, or some other factor at the moment unconsidered. On this question, I agree with Lave and Seskin (12), who reviewed the health costs of air pollution in the United States. They make the point that although such explanations may be shown in the future to have some basis, until such evidence is forthcoming, we have to attribute the differences that are being described in disease morbidity to the considerable differences in air pollution that exist in different regions. As a chest physician who has grown up during the period of discussion of the relationship between cigarettes and lung cancer, I think I should stress that the burden of proof of hitherto unrecognized factors lies with those who propose their existence.

Two New Concepts of Lung Function

At the same time as these epidemiologic studies were being planned and executed, significant advances were being made in two fields. In my view, these are becoming very relevant to any discussion of the implications of the epidemiologic data. The first has to do with the small airways of the lung as a primary site of abnormality, and the second is concerned with the defenses of the lung against the release of proteolytic enzymes, which probably come from cells within the lung performing their normal function.

Small airway function: It has become increasingly clear that although the small airways of the lung, perhaps those between 1 mm and 3 mm in diameter, contribute little *in toto* to the airway resistance of the lung as a whole, and for this reason may suffer major abnormality without such measurements as the FEV_1 being much affected (13) yet these airways appear to be a most important site for early change within the lung. Dr. Milic-Emili, Dr. Anthonisen, Dr.

20

Macklem, and other research workers have been studying the consequences of abnormalities in this part of the lung. In an interesting series of experiments, it has been proved possible to measure the volume at which airway closure occurs (14, 15). Earlier work had shown that the distribution of an inspired breath was determined in large part by whether it was taken from functional residual capacity begun from residual volume (16). Later work revealed that the explanation lay almost certainly in the relationship between the vertical height of the lung and the difference in intrapleural pressure this occasioned, related to the elastic recoil being exerted by the lung itself. It became apparent that one of the consequences of ageing in the lung would be that as the lung lost its elastic recoil, more and more airways at the most dependent part of the lung in the gravitational field would be closed at a higher and higher lung volume (17). It seems now quite clear that the lowering of arterial oxygen tension that occurs with age measured in recumbent patients is mainly a function of airway closure occurring in this situation. It has proved possible to explain the differences in gas distribution in the lung in relation to age entirely on this concept, and to relate the airway opening phenomenon in older people to the relationship that Edelman and co-workers found existed between the gas distribution in older people and the magnitude of the tidal volume (18). More recently, this concept has been taken a significant step further by the demonstration by Dr. Anthonisen and Dr. Martin that one can demonstrate abnormalities in time constants in small airways in 25-year-old cigarette smokers all with a normal FEV_1 and symptom free compared with subjects of the same age who are not cigarette smokers (19). This very sensitive technique indicates clearly, and perhaps for the first time, that the first site of effect of the inhalation of cigarette smoke is almost certainly at the terminal bronchiolar level. More recently still, Dr. Milic-Emili and Dr. McCarthy, doing work at Hammersmith Hospital, were able to show that the closing volume of small airways expressed as a per cent of vital capacity was abnormal in two thirds of

cigarette smokers in a group of 43 studied whereas the FEV_1 was significantly abnormal in only 11.6 per cent of the total group. A condensation of their recent studies is shown in figure 1, from which other deductions can also be made. Although the closing volume is slightly greater in cigarette smokers with symptoms, it was significantly abnormal in 8 smokers who had no symptoms and normal in 5 reasonably heavy smokers who did have symptoms of chronic bronchitis. I want to use these data to argue that they necessitate thinking about the site of effect of an irritant in much more discriminating terms than we have used in the past. Both of these methods are more sensitive than the FEV_1 because they are testing an abnormality in a different region of the lung. At this point, I might also add that the consequence of an abnormality of time constant in the small airways of the lung is that the compliance of the lung, when measured under dynamic circumstances, will have become frequency dependent, and Macklem and Woolcock showed some years ago that this was a feature of patients with established chronic bronchitis even if their airway resistance was normal or only marginally increased (20). It has also been shown that before the FEV_1 has become impaired in chronic bronchitis, there are disturbances of gas exchange, almost certainly due to nonuniformity of ventilation/perfusion ratios, in the lung periphery. The technique used to measure this with xenon-133 is not a simple procedure (21), but the results have been confirmed by Strieder and Kazemi by careful measurement of alveolar-arterial oxygen tension differences (22).

I believe that this body of data, taken together, gives us a new perspective on the problem of air pollutants and the lung as a whole, and it is for this reason that I have emphasized it. To me, it seems highly likely that if the small airways are sensitive to cigarette smoke or gases in cigarette smoke and are the first region to display measurable abnormality in cigarette smokers, that they may also be affected early by ambient pollutants.

Proteolytic enzymes in the lung: The other important area of recent advance is one

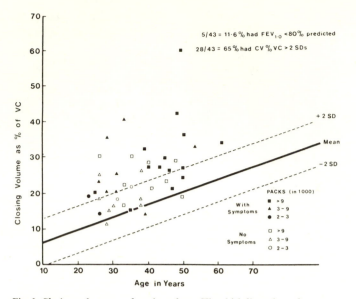

Fig. 1. Closing volume as a function of age. The thick lines show the regression between the measured lung volume at which airway closure can be demonstrated to have occurred and age in a nonsmoking population with two standard deviations (broken lines) on either side of this mean. Some cigarette smokers with and without symptoms of bronchitis and even at a young age had a closing volume substantially higher than normal. In this group of 43 subjects, 28 had a closing volume more than two standard deviations greater than a noncigarette smoking population, whereas only 5 had an FEV_1 that was less than 80 per cent of predicted. (Data from Dr. Milic-Emili and Dr. McCarthy, Royal Postgraduate Hospital, Hammersmith).

in which I have not been working, and therefore, my account of it will have to be even briefer than that which I have devoted to other sections. The discovery by Eriksson in Sweden (23) that an inherited deficiency of serum α_1-antitrypsin predisposed the patient to alveolar destruction in the absence of chronic bronchitis was unquestionably of the greatest general significance, even if most patients we see with emphysema do not have particularly low concentrations of serum α_1-antitrypsin. Later work has shown that the lung is almost certainly protected from proteolytic enzymes by a series of antiproteolytic enzymes, of which α_1-antitrypsin is one. It seems clear that the release of proteolytic enzymes occurs from cells, possibly principally macrophages, that are responsible for cleaning of the surfaces of the alveoli and bronchioles and removing foreign material and bacteria from these regions (24). A wide variety of different challenges, both particulate and gaseous, and particularly probably both of these together, lead to an increase of cells and macrophages in the lung as a more or less inevitable consequence either of direct irritation or of deposition in the alveoli of foreign material (25). It seems a very fair assumption that anything that increases the cellular content of the lung would also ultimately increase the release of proteolytic enzymes likely to occur within it, and hence, the defenses against such proteolytic enzymes become of great importance. Only if these are reasonably good will the alveoli be protected from destruction. There are many other aspects of this work currently in train, and there is without doubt much more to be learned about the general pro-

22

cesses involved.

The type of emphysema induced by papain administered intratracheally to dogs looks like human panlobular emphysema. The internal surface area is reduced, and the mean linear intercept of the lung is increased. Dr. Macklem and Dr. Thurlbeck and their colleagues have recently shown quite convincingly that this morphologic lesion leads to a significant reduction in diffusing capacity and elastic recoil and an increase in total lung capacity, functional residual capacity, and residual volume (26). There seems good reason to suggest that the presence of morphologic emphysema of this kind of itself is capable of giving rise to the characteristic pattern of functional change that is associated with emphysema, and although the analogy should not be pressed too far, there seems no reason to feel that an element of chronic inflammatory change in major bronchi is a necessary constituent of this syndrome.

If animals are given particles of a size to get into alveoli and together with these or separately from them are also exposed to an irritant gas such as oxides of nitrogen, then the damage caused to the lung appears to be greater if both are given together than if each animal is only given one of them (27), and if the challenges are delivered in an alternating way, the damage is also potentiated. This suggests that the defenses evoked by one of the challenges may impair the protection that the lung has against the other. These observations, which admittedly have been made on occasion with greater concentrations of pollutants than normally encountered, should at least alert us to the possibility that a certain ambient particle concentration might be without hazard in an otherwise clean environment, but harmful if respired in association with gases commonly found in an urban environment.

It is also clear from animal experiments that exposure to one challenge, such as ozone, may impair the animal's defenses against infection (28), and this effect of ozone may be demonstrated at the same concentrations to which some populations are now being exposed to the gas. In animals, it has also been shown that this pollutant has a protective effect on radiation damage to the lung, and we certainly have to consider that it may be that certain influences exert a protective effect in relation to others.

Differential Sites of Action

The main danger of any attempt to synthesize is that the oversimplification required ends by producing misleading statements. To illustrate the observation that a single pollutant may act at distinct sites in the lung, the example of ozone is a suitable prototype. This gas is formed by the action of sunlight on oxides of nitrogen produced by automobiles in the presence of hydrocarbons. In the suburbs of Los Angeles, concentrations of ozone reach 0.70 ppm for brief periods, usually during the lunch hour and mostly in the suburbs of Los Angeles (29). Accordingly, we have been studying the effect of 0.70 ppm of ozone in the absence of all other pollutants on normal subjects (30). A synopsis of these data in 9 normal subjects who sat at rest in such an atmosphere of ozone is presented in table 1. All of these except one noticed soreness of the upper airway and a tendency to cough when taking a deep breath. The most significantly disturbed measurement was that the maximal negative intrapleural pressure (P_L max) that they could exert after a two-hour exposure had diminished by 4.4 cm of water. This effect is presumably explained by stimulation of irritant receptors in the upper airway that inhibits the taking of a deep breath. The physical properties of the lung, as shown by its static compliance, were unchanged, indicating that this difference in maximal negative pressure had nothing to do with a change in physical state of the lung. Also, after two hours' exposure at rest, there had been a just significant increase in airway resistance (R_L) but no change in the fractional carbon monoxide uptake. The dynamic compliance was lower at all frequencies, but with this period of exposure the results just failed to reach significance. This study illustrates that this gas breathed in this way without mouthpiece and noseclip but in a chamber for two hours produces its effects mainly in the upper airway, stimulating irritant receptors and inhibiting the taking of a deep breath. The gas is not exerting much effect at the terminal bronchiolar level although the fact that the dynamic compliance was regularly lower at all

23

TABLE 1

NORMAL SUBJECTS EXPOSED TO 0.70
PPM OF OZONE AT REST

(n = 9)	At Start	After Two Hours	
Vital capacity, liter	6.2	6.0	NS
P_L max, cm H_2O	41.3	36.9	$P < 0.005$
C_{st} deflation, liter/cm H_2O	0.297	0.302	NS
R_L at normal frequency	1.44	1.76	$P < 0.025$
%CO Uptake	0.577	0.556	NS

Although C_{dyn} was lower after ozone exposure at all frequencies, results did not reach significance. *Definition of symbols*: P_L max = maximal intrapleural pressure; R_L = increase in airway resistance; C_{st} = static compliance; C_{dyn} = dynamic compliance.

frequencies indicates that it must have begun to have an effect at this site. The results in 3 subjects who during the period of exposure had alternating periods of rest with periods of 15 minutes of light exercise on a bicycle ergometer are shown in table 2. The degree of exercise was sufficient to double the ventilation. One of these 3 subjects (Subject B) was completely resistant to this concentration of ozone and was normal throughout. In Subject A, a marked decrease in P_L max occurred, the airway resistance was significantly increased, and the dynamic compliance at the end of two hours was one half of the value it was at the beginning, which is a highly significant change. The fractional carbon monoxide (CO) uptake did not alter, indicating that no effect was being exerted at the alveolar level nor on the ventilation/perfusion distribution within the lung. In Subject C, not only was there the largest decrease in P_L max, but the dynamic compliance was also greatly reduced, and the CO uptake was significantly diminished after the two hours of exposure. In this subject, therefore, ozone not only had a major effect in the upper airway, but it now produced a reduced dynamic compliance indicating nonuniform time constants in small bronchioles and in addition produced a considerable impairment of fractional CO uptake. I do not want to talk particularly about ozone as a pollutant, but I want to use these data to illustrate that this same gas may, at a given concentration, and

depending on the time during which it is breathed and the exercise level of the subject, produce irritant effects in the upper airway without any other change, or it may impair the time constants in terminal bronchioles, or produce changes in gas exchange indicating either an alveolar effect or a disturbance of ventilation/perfusion ratio within the lung. The considerable individual variation among subjects in the magnitude of effects is a commonplace observation. It is always tempting to talk of an irritant gas as if its effects were just those of bronchospasm, but obviously our thinking in relation to the sites of action of a gas such as ozone must be much more sophisticated than to confine our attention to only one particular phenomenon. The edemagenic properties of ozone in greater concentrations have been known for a long time, and it is also known that exposure of animals to ozone causes an increase in the number of cells within the lung; therefore, to the acute effects of ozone that we have been studying, one must add the possibility of undesirable long-term effects. Ozone acts on materials such as rubber to alter the linkages within it and impair its elasticity. One can, in fact, more easily set acceptable upper concentrations of ozone for a population that wishes to undertake light exercise in such an environment for a period of two hours, than one can be sure that the population is adequately protected against possible

TABLE 2

EXPOSURE OF NORMAL SUBJECTS TO
0.70 PPM OF OZONE WITH INTERMITTENT
LIGHT EXERCISE*

(n = 3)	At Start	After Two Hours
SUBJECT A		
P_L max, cm H_2O	39.0	30.0
R_L, cm H_2O/liter/sec	1.4	2.4
%CO Uptake	0.613	0.593
C_{dyn}, liter/cm H_2O	0.26	0.13
SUBJECT B No change observed in any parameter		
SUBJECT C		
P_L max, cm H_2O	47.0	35.0
R_L, cm H_2O/liter/sec	1.6	1.9
%CO Uptake	0.598	0.415
C_{dyn}, liter/cm H_2O	0.128	0.078

*The bicycle exercise was sufficient to double the ventilation. For definitions of symbols, see table 1.

long-term effects of a gas of this kind. For similar reasons, the existing data on sulfur dioxide exposure as it relates to normal human subjects are highly unsatisfactory. Not only have relatively unsophisticated methods been used to measure the effect on the lung, but the periods of exposure have often been quite short. Furthermore, very few studies have been done with subjects exercising normally, as they do if they walk along a city street for example, and very little attempt has yet been made to see if the effect of sulfur dioxide can be traced down the airway in a similar fashion to what appears to be the case for ozone. Neither of these gases has yet been studied with new techniques of measuring airway closure, and it may be that this technique of studying pulmonary function will give us much better indications of early effects in small airways than any methods we have used in the past.

I mentioned that this review of existing data might lead us to see more clearly where our deficiencies in knowledge had to be made up, and it is obvious from a review of existing data on the acute effect of oxides of nitrogen, sulfur dioxide, and ozone in human subjects that a great deal of work lies ahead of us. Even when acute effects are reasonably well understood, we have to deal with long-term effects, which are of course much more difficult to study. The point I wish to make at this stage is that we are not yet in possession of a satisfactory body of knowledge on the acute effects.

Three Major Sites of Action

We are now in a position to try to differentiate in a very preliminary way the effects that any inhaled substance may have on the human lung. First, the effects can be examined in terms of the site of action of the material. We perhaps would do well to try to make some preliminary statements on the possible or documented effects of different materials on the major bronchi of the lung, on the terminal bronchioles, and on the alveoli themselves. These are shown in diagrammatic form in figures 2, 3, and 4. The first effect of irritants in major bronchi may be to inhibit the taking of a deep breath through the stimulation of the irritant re-

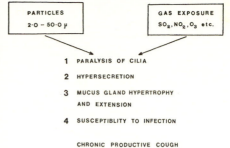

Fig. 2. Effect of irritants in major bronchi.

ceptors, a phenomenon that the data with ozone demonstrate. Presumably, reversible bronchospasm will also be caused, and in both of those effects, there will be considerable individual variation. The longer term effects are probably more important, and as shown in figure 2, these consist of paralysis of cilia, hypersecretion of bronchial mucous glands, and mucous gland hypertrophy and extension into smaller airways, a phenomenon documented by Reid to occur as a result of exposure of rats to high concentration of sulfur dioxide (31). One may note that SO_2 has a predilection for affecting the major bronchi, the acute inhalation of very high concentrations in man leading to generalized bronchiectasis (32). An increased susceptibility to infection might be a consequence of any of these latter changes, and they might also lead to a chronic productive cough in the long term. One must assume that particles of a size favorable to deposition in this part of the airway as well as gases might have some or all of these effects, but the ciliary system, if functioning, should deal adequately with large particles. One can argue that all the phenomena in relation to the epidemiologic data could be explained by the action of irritants at this site. Thus, one can hypothesize that the decreased FEV_1 demonstrated by Holland and Reid (6) in middle-aged men in the more polluted zones (amounting to a decrement of 0.3 liters in moderate cigarette smokers) was in fact due to more hypersecretion, more mucous gland hypertrophy, and more repetitive infection, and the result of this was an increase in upper airway resistance as shown by a diminution in FEV_1.

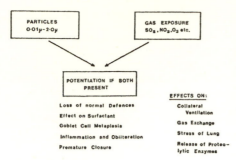

Fig. 3. Effect of irritants in terminal bronchioles.

Similarly, the increased occurrence of lower chest infection in infants might be ascribed to these mechanisms.

In figure 3 are shown possible effects in terminal bronchioles. Smaller particles will penetrate and deposit in this region, and there is a particular reason to stress the possible potentiation of effect if gases and particles are inhaled concurrently as in cigarette smoke. Most of the sulfur dioxide that is breathed in is stopped in the nose, and soluble gases such as oxides of nitrogen similarly are diminished if breathed in the gas form. Combined with particles, there is a reasonable chance that these will get further down the lung and possibly have a greater effect in lower concentration. There are also experimental data and some scattered human data indicating that acute exposure to high concentrations of oxides of nitrogen characteristically produce pulmonary edema and fibrosing bronchiolitis rather than chronic bronchitis or bronchiectasis (33).

Is it a coincidence that the first effects of cigarette smoke are in small airways, and that cigarette smoke contains high concentrations of oxides of nitrogen? What may occur in terminal bronchioles as a result of these gas exposures? Again one can argue in general that the defenses will be lost; there may be an adverse effect on surfactant, which may have something to do with keeping these small airways patent; there may be metaplasia of goblet cells and inflammation and obliteration. Any or all of these phenomena will lead to premature closure of airways, and this is presumed to be what is

being demonstrated with the increased closing volume demonstrated in cigarette smokers (figure 1). Cigarette smoking even at a young age and for only relatively few years and in the absence of symptoms of bronchitis, can by virtue of these effects lead to premature closure of airways. If there is much cellular infiltration around terminal bronchioles as a consequence of these challenges, we may expect some release of proteolytic enzymes in the region, but we do not have much idea of what may result from this phenomenon. In interpreting this chain of events, we have to take particular note of the fact that many terminal bronchioles may be compromised or lost before the FEV_1 is much affected. Although the terminal bronchioles appear to be the first part of the lung to be compromised in cigarette smoking, there is as yet no direct experimental evidence on which to base an argument that it is similarly the first point of attack in terms of air pollutants. What kind of experiments would be needed to demonstrate this? One would need to study the airway closing volume of the noncigarette smoking population in areas of different air pollution, together with the frequency dependence of compliance in the same population. There is no reason why the early adverse effect of air pollutants, if confined to

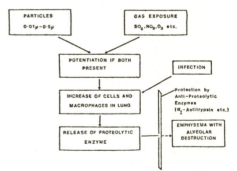

Fig. 4. Effect of irritants in alveoli.

the terminal bronchiolar region, would be demonstrable by symptom surveys, radiographic change, or demonstrated differences in FEV_1, because in cigarette smokers, quite striking effects have been shown to be present in this particular region of the lung with none of those being sensitive indicators of them. The premature airway closure would have important consequences as the lung ages, since with progressive loss of lung recoil, the consequences of premature airway closure would become more severe (17).

The effect of irritants in the alveoli are different from either of these (figure 4). Firstly, particles of the range order 0.01 μ to 0.05 μ are particularly likely to get to this region (34), and in modern city air there are plenty of particles in this size range. We have no way of calculating how much of a gas such as ozone in fact reaches the alveolar level because it is so reactive to surfaces, and one imagines that much of it has disappeared before reaching the alveolar wall. Some gases of the nature of sulfur dioxide may be absorbed in the nose and later removed through the alveoli and exert their effect not by having been delivered to the alveoli through the bronchial tree but by rediffusion out from the blood into alveolar gas. There is evidence that there is potentiation of effect if both particles and gases are present, and one major consequence of any of these phenomena is an increase of cells and macrophages within the lung. There is plenty of evidence that this is the primary response to most of these challenges, and an increased aggregation of macrophages and leukocytes occurs quite quickly in response to most of them (25). One associates an increased cellular content of the lung primarily with infection, and this of course also will have a major role though probably an intermittent one in increasing the cellularity of the lung. The presence of these cells will increase the chances of release of proteolytic enzymes, and as pointed out earlier, if the defenses against these are not good, we have the possibility of emphysema developing with alveolar destruction. The release of proteolytic enzymes therefore represents a threat to the lung that it may have some difficulty meeting if it is not adequately provided with antiproteolytic enzymes.

There is an additional possible consequence of gases at the alveolar level that they may have altered the physical characteristics of the lung in the same way as occurs in normal ageing. As the lung ages, elastic recoil is continuously diminished although the morphologic loss of alveoli is quite small (35). No one knows why this occurs nor what rearrangement of collagen or elastin may be occurring to produce it. Yet that it occurs naturally and that the lung is a huge organ continuously exposed to the outside environment, naturally raises the question as to whether this ageing is not in fact a consequence of environmental agents in the first place. Gases such as ozone might well have an effect, accelerating what is essentially a normal process, and hence accelerating the loss of recoil that would normally be ascribed to age. This chain of reasoning immediately leads to the question of whether the pressure-volume curves of the lung of the nonsmoking population exposed to ozone in places such as Pasadena is any different from the same curves in the nonsmoking population living in cities that do not have a problem with oxidant air pollution. As far as I know, no data have been collected relating to this problem although the urgent need for it has been evident for some years.

As noted above, one might explain the data of Holland and Reid (6) solely on a basis of increased large airway resistance. On the other hand, it may just as reasonably be argued that the main cause of their observed difference is a greater extent of morphologic emphysema in men and in the more polluted region. From their data, the FEV_1 for a man smoking between 1 g and 14 g of tobacco per day, if he lived in a country district, was 3.0 liters. If he lived in London, it was approximately 2.7 liters. This decrement of FEV_1 of 0.3 liters, if looked at in terms of the relationship between the FEV_1 and the extent of morphologic emphysema (36), would result from a quite small increment in morphologic emphysema, approximately an increase of only 2 of 30 units of morphologic destruction. This difference would be hard to measure precisely even with modern quantitative techniques. This is, of course, not to say that the decrement of function Holland and Reid observed was caused by such a difference in the extent of morphologic emphysema, but only that the significant difference in pulmonary function that they documented could have resulted from a difference in morphologic emphysema so small that the pathologist would find it difficult to demonstrate.

Present Problems and Conclusions

In trying to bring together information of this kind, there are several inherent difficulties. First, just as railway lines appear to be approaching each other as they recede into the distance, one may be deceived in looking forward into seeing apparent convergencies that do not in reality exist. Equally, one may be inhibited from thinking in sufficiently broad terms on such a problem and confine oneself to making such remarks, as one quite often hears, as that "there are only one or two ways in which the lung can respond to a physical challenge." Too much emphasis on the latter point of view leads one to express every effect in terms of one condition rather empirically defined, such as chronic bronchitis. In trying to bring this material together, therefore, I have to try to avoid both of these potential difficulties.

In reviewing the present status of knowledge in these fields, I come to four general conclusions that seem to me to be of importance.

First, we must make a conscious effort to think of the effect of any given pollutant in at least three major areas of the lung, and we must in no sense restrict our consideration of its effect to any one of these until we are reasonably sure that other regions are not involved. To assume that the only major importance of particle pollution is to cause some reversible bronchospasm, or that the only thing ozone does is to irritate receptors in the upper bronchial tree, is clearly much too simplistic a viewpoint to sustain. It is important to develop the custom of thinking about possible effects at three different levels in the lung and of trying to think of such effects as bronchospasm, mucous gland hypertrophy, obliteration of terminal bronchioles, and increased cellularity, all at once rather than individually or, what is worse, in different types of scientific journals.

Second, I have drawn attention to the observations indicating clearly that the earliest effects of cigarette smoke occur in small airways, that the potentiation of chest infections that occurs in 20-year-old cigarette smokers indicates that interference at this level may impair lung defenses against infection, that the small airways are a major if not the principal attack site in early chronic bronchitis, and that the FEV_1 and the chest film are too crude indicators of change in this region of the lung to be of much use to us. We do not know whether cigarette smoke is a good model for air pollution, but we can recognize that if the earliest effects of some air pollutants are exerted at the level of small airways, the questionnaire, the chest film, and the FEV_1 are much too crude as tools to indicate such effects.

Third, I have pointed out that we must be on guard against long-term effects going undetected. Most pollutants increase lung cellularity, and this increases the hazard of proteolytic enzyme release, and this in turn may relate to alveolar loss or emphysema. It is certainly possible that some pollutants, perhaps particularly ozone, impair lung recoil and accelerate ageing with no major effects on the airway.

Fourth, it is hardly necessary to emphasize that a great deal of work remains to be done. I will be satisfied if what I have said persuades some of you who were not previously convinced that we are faced with a major problem in respect to air pollution and its effect on the human lung. Ten years ago, it was popular to remark that although we had air pollution, we had no disease to go along with it. We now know that increased respiratory disease morbidity is almost certainly related to it. No one will dispute the fact that cigarettes have been responsible for a far greater burden of chronic respiratory disease than air pollution, but we have to start working now towards some of the goals I have mentioned if we are going to play an effective part in setting comparative priorities in respect to air pollution control. An examination of the effects of air pollutants and the epidemiologic data along the lines that I have attempted indicates clearly the need for more thorough studies of the state of the lung in the noncigarette smoking population exposed to different air contaminants, and I very much hope that in the next few years some data of that kind have been secured. The general level of "production" of oxides of nitrogen for the United States from automobiles is predicted to increase from approximately 7×10^6 tons per year in 1970 to approximately 17×10^6 per year in 1990 (37). No responsible physician, and certainly no responsible chest physician, having reviewed the data, can look upon that possibility with equanimity.

It is my hope that by bringing to light some of the important and unanswered questions in this field, I may stimulate one or two clinical investigators and physicians to turn their attention to some part of it. It is unfortunate that contemporary scientific society pays little attention to applied work of this kind, apparently having convinced itself that it is an applied area of research only suitable for those of inferior intellect. Some very distinguished work has been done in relation to epidemiology and air pollution and in studying such problems as particle deposition in the lung, but I have not noticed the scientific community paying very much attention to it. I can safely make these remarks because I do not work in either of those fields myself. I hope that by indicating some of the areas of interaction and present interest, I may have stimulated one or two people to devote their scientific talents to these questions. I hope that by the time they have achieved success, the scientific community may, by a slight reorientation of its objectives, give good recognition to the contribution they will have made.

Acknowledgments

I wish to thank Dr. Francis D. North whose generosity has endowed this lectureship; my colleagues, Dr. Nick Anthonisen, Dr. Peter Macklem, and Dr. Milic-Emili for allowing me to use their data and for assisting me to keep abreast of recent advances; and the many students and members of the public who by their concern for the future environment have greatly accelerated a necessary reorientation of thinking in relation to air pollution.

References

1. Mortality and morbidity during the London fog of December 1952, Ministry of Health Report on Public Health No. 95, London, Her Majesty's Stationery Office, 1954.
2. Lawther, P. J.: Climate, air pollution and chronic bronchitis, Proc. Roy. Soc. Med., 1958, *51*, 262.
3. Air pollution and health. A Report for the Royal College of Physicians of London, Pitman Medical and Scientific Publishing Co., Ltd., London, 1970, p. 80.
4. Eisenbud, M.: Environmental protection in the city of New York, Science, 1970, *170*, 706.
5. Petrilli, F. L., Agnese, G., and Kanitz, S.: Epidemiologic studies of air pollution effects in Genoa, Italy, Arch. Environ. Health (Chicago), 1966, *12*, 733.
6. Holland, W. W., and Reid, D. D.: Urban factor in chronic bronchitis, Lancet, 1965, *1*, 445.
7. Lambert, P. M., and Reid, D. D.: Smoking, air pollution, and bronchitis in Britain, Lancet, 1970, *1*, 853.
8. Douglas, J. W. B., and Waller, R. E.: Air pollution and respiratory infection in children, Brit. J. Prev. Soc. Med., 1966, *20*, 1.
9. Air quality criteria for nitrogen oxides, Air Pollution Control Office, Publication No. AP-84, Environmental Protection Agency, Washington, D.C., 1971.
10. Toyama, T.: Air pollution and its health effects in Japan, Arch. Environ. Health (Chicago), 1964, *8*, 153.
11. Colley, J. R. T., and Reid, D. D.: Urban and social origins of childhood bronchitis in England and Wales, Brit. Med. J., 1970, *2*, 213.
12. Lave, L. B., and Seskin, E. P.: Air pollution and human health, Science, 1970, *169*, 723.
13. Macklem, P. T., and Mead, J.: Resistance of central and peripheral airways measured by retrograde catheter, J. Appl. Physiol., 1967, *22*, 395.
14. Simonsson, B. G.: Clinical and physiological studies on chronic bronchitis: II. Classification according to spirometric findings and effect of bronchodilators, Acta Allerg. (Kobenhavn.), 1965, *20*, 301.
15. Leblanc, P., Ruff, F., and Milic-Emili, J.: Effects of age and body position on "airway closure" in man, J. Appl. Physiol., 1970, *28*, 448.
16. Dollfuss, R. E., Milic-Emili, J., and Bates, D. V.: Regional ventilation of the lung studied with boluses of xenon[133], Resp. Physiol., 1967, *2*, 234.
17. Holland, J., Milic-Emili, J., Macklem, P. T., and Bates, D. V.: Regional distribution of pulmonary ventilation and perfusion in elderly subjects, J. Clin. Invest., 1968, *47*, 81.
18. Edelman, N. H., Mittman, C., Norris, A. H., and Shock, N. W.: Effects of respiratory pat-

tern on age differences in ventilation uniformity, J. Appl. Physiol., 1968, 24, 49.

19. Martin, R. R., and Anthonisen, N. R.: Frequency dependence of intrapulmonary distribution of 133Xe, Program of Amer. Fed. for Clin. Res., Atlantic City, 1971, Abstract 215, p. 64a.

20. Woolcock, A. J., Vincent, N. J., and Macklem, P. T.: Frequency dependence of compliance as a test for obstruction in small airways, J. Clin. Invest., 1969, 48, 1097.

21. Anthonisen, N. R., Bass, H., Oriol, A., Place, R. E. G., and Bates, D. V.: Regional lung function in patients with chronic bronchitis, Clin. Sci., 1968, 35, 495.

22. Strieder, D. J., and Kazemi, H.: Hypoxemia in young asymptomatic cigarette smokers, Ann. Thorac. Surg., 1967, 4, 523.

23. Eriksson, S.: Studies in α_1-antitrypsin deficiency, Acta. Med. Scand., 1965, 177 (Supplement 432, p. 1).

24. Green, G. M.: Integrated defense mechanisms in models of chronic pulmonary disease, Arch. Intern. Med. (Chicago), 1970, 126, 500.

25. Brain, J. D.: Free cells in the lungs, Arch. Intern. Med. (Chicago), 1970, 126, 477.

26. Pushpakom, R., Hogg, J. C., Woolcock, A. J., Angus, A. E., Macklem, P. T., and Thurlbeck, W. M.: Experimental papain-induced emphysema in Dogs, Amer. Rev. Resp. Dis., 1970, 102, 778.

27. Boren, H. G.: Pathobiology of air pollutants, Environ. Res., 1967, 1, 178.

28. Air quality criteria for photochemical oxidants, National Air Pollution Control Administration Publication No. AP-63, Washington, D. C., 1970.

29. Mosher, J. C., Macbeth, W. G., Leonard, M. J., Mullins, T. P., and Brunelle, M. F.: The distribution of contaminants in the Los Angeles Basin resulting from atmospheric reactions and transport, J. Air Pollut. Contr. Assn., 1970, 20, 35.

30. Bates, D. V., Bell, G., Burnham, C., Hazuvha, M., Mantha, J., Pengelly, L. D., and Silverman, F.: Problems in studies of human exposure to air pollutants, Canad. Med. Assn. J., 1970, 103, 833.

31. Reid, L.: Evaluation of model systems for study of airway epithelium, cilia, and mucus, Arch. Intern. Med. (Chicago), 1970, 126, 428.

32. Bates, D. V., and Christie, R. V.: Respiratory function in disease, W. B. Saunders Co., Philadelphia, 1964, p. 396.

33. Becklake, M. R., Goldman, H. I., Bosman, A. R., and Freed, C. C.: The long-term effects of exposure to nitrous fumes, Amer. Rev. Tuberc., 1957, 76, 398.

34. Deposition and retention models for internal dosimetry of the human respiratory tract, Report by Task Force on Lung Dynamics, Health Phys., 1966, 12, 173.

35. Turner, J. M., Mead, J., and Wohl, M. E.: Elasticity of human lungs in relation to age, J. Appl. Physiol., 1968, 25, 664.

36. Thurlbeck, W. M., Henderson, J. A. M., Fraser, R. G., and Bates, D. V.: Chronic obstructive lung disease: A comparison between clinical, roentgenologic, functional, and morphologic criteria in chronic bronchitis, emphysema, asthma, and bronchiectasis, Medicine (Balt.), 1970, 49, 81.

37. Nationwide inventory of air pollutant emissions, 1968, Environmental Health Service, National Air Pollution Control Administration Publication No. AP-73.

30

Air pollution

and

exacerbations of bronchitis

P. J. LAWTHER, R. E. WALLER,

and MAUREEN HENDERSON

In the notorious London fog of December 1952 the deaths of some 4,000 were attributed to the effects of the extraordinarily high concentrations of smoke, sulphur dioxide, and other pollutants reached at the time (Ministry of Health, 1954). Subsequently it was shown (Gore and Shaddick, 1958; Bradley, Logan, and Martin, 1958) that in other foggy periods in London changes in mortality could be related to increases in the concentrations of smoke and sulphur dioxide. More detailed studies revealed a general relationship between mortality and morbidity in Greater London and the concentrations of smoke and sulphur dioxide throughout the winter months (Martin and Bradley, 1960; Martin, 1961). Since these studies demonstrated that effects occurred in the general population, we have used a simple technique to follow the response of selected individuals to periods of high pollution. Although the data have been derived from subjective reports, the results have proved useful in assessing the relative importance of pollution and weather in producing exacerbations of bronchitis. From small beginnings in the winter of 1954–55, these studies were enlarged to involve approximately 1,000 patients both in 1959–60 and 1964–65, so as to follow the effects of changes in pollution arising out of clean air policies. Although further work is in progress, here we describe the gradual development of the technique and review the results to date.

31

In the winter of 1954–55 pocket diaries were issued to 34 patients attending the Emphysema Clinic at St. Bartholomew's Hospital. These patients were asked to record their own assessments of their state of health day by day by means of the following code:

A = condition better than usual
B = condition the same as usual
C = condition worse than usual
D = condition much worse than usual.

The method was kept as simple as possible, and the diaries used had only a small space per day, to encourage the patients to use the code rather than to write comments. The results were examined around a period of high pollution in January 1955 (Waller and Lawther, 1955) and supported the view that high concentrations of pollution, measured in terms of smoke, sulphur dioxide or other associated pollutants, were deleterious to the health of bronchitic patients, even in the absence of wet fog.

In the following winter the study was extended to include 195 patients attending four other centres in London (the Hammersmith Hospital and the Chest Clinics at East Ham, Croydon, and Edgware). All of these patients had a history of bronchitis, 80% of them were men, and the average age of the group was 56 years (range 27 to 78 years). Other groups were enrolled in Sheffield (85 patients), in Manchester (35 patients), and in the West Midlands (19 patients).

The diary code was extended so that the patients in addition could indicate:

F = fog at some time during day
X = head or chest cold that day
H = indoors all day.

When these codes applied, some patients omitted to enter A, B, C or D in their diaries, so this departure from simplicity impaired the completeness of reporting.

Replicated sheets were used instead of diaries for the first two months, and cheap pocket diaries were issued to all the patients on 1 January 1956. The age, occupation, and clinical diagnosis were noted for each patient. The results for this winter were assessed by means of an arbitrary scoring system:

A (better) = −1   C (worse)   = 1
B (same)   = 0   D (much worse) = 2

The mean score for each day was determined

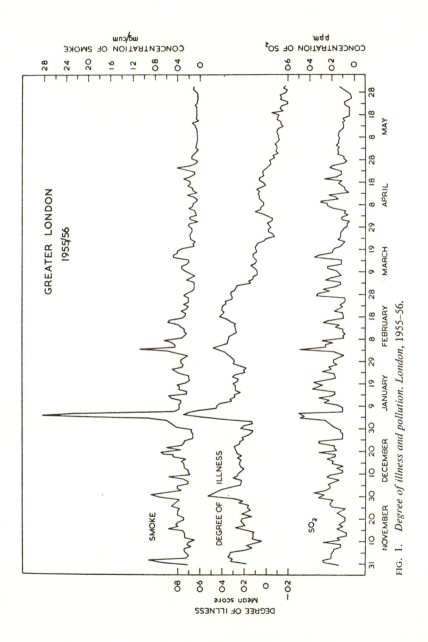

FIG. 1. *Degree of illness and pollution. London, 1955–56.*

from the sum of the individual scores divided by the total number of entries for the day. Approximately 180 London patients filled in their diaries regularly: the results were assessed separately for each of the five centres, but finally they were pooled.

RESULTS From November to February there was a general increase in the degree of illness of the group and the several sharp peaks superimposed on this trend coincided with increases in the concentrations of smoke and sulphur dioxide as measured at the sampling site outside our own laboratory in the centre of London (Fig. 1). There was no consistent relationship with temperature or humidity measured at the Meteorological Office, close to our laboratory, although from March to May, when there was an overall decline in the degree of illness, there were a few minor peaks in the illness curve that coincided with falls in temperature (Fig. 2). The association between degree of illness and pollution could have been fortuitous if patients who expected to be worse in 'fog' merely entered C or D in their diaries when they noticed that visibility was reduced. There was some association between visibility and degree of illness (Fig. 3), but there was little response on the day of lowest visibility (19 December 1955) when wet fog was present but the concentration of smoke was not exceptionally high. Most of the minor changes in pollution produced changes in visibility that were too small to be recognized, so we thought it likely that the patients' entries indicated real changes in condition in relation to pollution (Waller and Lawther, 1957).

The condition of the London patients changed uniformly, irrespective of their area of residence. Although the average concentrations of pollution differed from one area to another, increases and decreases occurred at about the same time everywhere. The situation was different in Sheffield, where with many hills and valleys it was possible for pollution to accumulate in some areas and not others. Although there was a consistent response to the major episode of high pollution in early January in all the cities in this study, the association between illness and pollution was not in general as clear cut in Sheffield as in London. Later, more detailed studies of the effects of pollution were undertaken in Sheffield (Clifton, 1967).

In Manchester (Fig. 4), the pattern of response

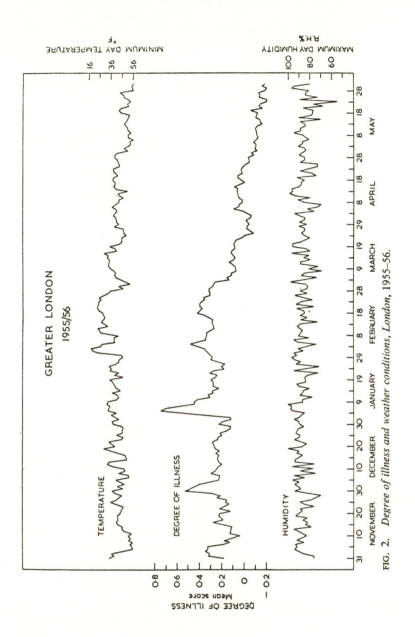

FIG. 2. *Degree of illness and weather conditions, London, 1955–56.*

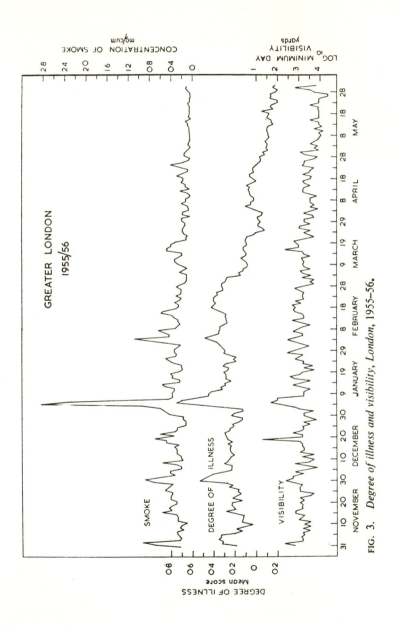

FIG. 3. *Degree of illness and visibility, London, 1955–56.*

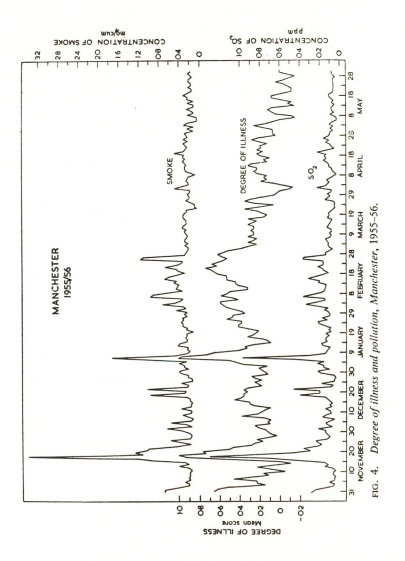

FIG. 4. *Degree of illness and pollution, Manchester, 1955–56.*

37

was similar to that found in London, with a big change in the degree of illness on three occasions when pollution was high (at the end of October, in November, and in early January). There were not many patients in the West Midlands group and the random variation was large, but there was evidence of a response to high pollution on two or three occasions during the winter.

The interpretation of our results was based mainly on inspection of composite graphs which showed little or no time lag between increases in pollution and increases in the degree of illness. However, the exact time at which pollution increased in any given locality could not be established, neither could the time of day when patients began to feel worse. The pollution measurements were average concentrations over periods of 24 hours beginning and ending at noon, and the patients' diary entries referred to their overall assessment of condition each day, beginning and ending nominally at midnight. In Figs 1 to 4 the time scales have been matched as closely as possible by plotting all results at the mid-points of the periods concerned, but even then some latitude must be allowed in matching up peaks in pollution and illness. Hourly measurements made at our laboratory showed that most 'episodes' of high pollution had sharp peaks in concentrations of smoke and sulphur dioxide, lasting for a few hours (Waller and Commins, 1966), and if the patients were influenced more by maximum values than by 24-hour average concentrations, the 'stimulus' and the 'response' could appear to be out of phase by half a day either way. During periods of high pollution adverse conditions often spread slowly across London, so that patients in some areas might have been affected later than others. Despite these uncertainties about times, there was little doubt about the association between illness and pollution, and in general when the concentrations of smoke and sulphur dioxide increased suddenly, there was a rapid rise in the degree of illness, followed by a more gradual return to normal. This type of pattern made it difficult to assess the association adequately in terms of correlation coefficients, since pollution was often low again before the patients had recovered, and there was also the problem of allowing for the gradual change in 'baseline' in the degree of illness figures during the winter. Some idea of the relative importance of these variables was obtained by calculating correlation coefficients between each of them and the mean score, for each month separately. In London, mean score was significantly correlated with both smoke and sulphur dioxide for most of the winter, but in the other centres, where the random variation was greater, the coefficients were significant only for one or two months. In general the correlation with temperature and relative humidity was small and not significant. The results suggested that the concentrations of either smoke or sulphur dioxide could be used as indices of adverse conditions, but they gave no indication as to whether one of these pollutants was any more important than the other.

FURTHER STUDIES

Although these early studies left a number of questions unanswered, they showed that valuable information about the nature of the environmental conditions affecting bronchitic patients could be obtained by use of this simple technique, without any special care in the clinical selection of patients or location of sampling sites. Most of the patients maintained their interest and enthusiasm for filling in their diaries daily for one winter, but it began to wane once they were asked to repeat the experiment through a second winter.

No diaries were issued during the winter 1956–57, but in 1957–58 a further study was undertaken in London. Patients were again recruited from the clinics at the Hammersmith Hospital, East Ham, Croydon, Hammersmith, and Edgware, and a new centre (Enfield) was included. The aim was to recruit at least 50 patients from each centre, and in this study arrangements were made to measure pollution locally, where the subjects lived or worked, with a view to assessing the results from each area separately. At Croydon sampling apparatus was installed in the home of one of the patients; at Enfield it was installed in a factory; and at the remaining three centres it was installed in the Chest Clinic. The apparatus consisted of a standard daily smoke filter, as used at our own laboratory and at many other sites throughout the country (Warren Spring Laboratory, 1966), but sulphur dioxide was not measured. The smoke filters were changed daily by the occupants or staff at each site and sent to us once a week for assessment. Brief personal and clinical notes were obtained for each of the 246 patients originally enrolled for the 1957–58 study, and at some time during the course of the winter each was visited in his (or her) home. Nearly all the patients were within the age range 45 to 70: the mean age was 58 and 84% were male. Most were diagnosed as chronic bronchitics, with a history of cough and phlegm for more than five years. Few reported severe breathlessness, but most gave some history of wheezing. Just over half the men were in regular employment.

The type of heating in use was noted when visiting the homes: 71% had open coal fires, and a further 9% were using smokeless fuels on open fires. These proportions were typical of London homes at that time, but the situation is different now.

The results of the 1957–58 study were initially assessed on the 'scoring system', and as before there were sharp peaks in the illness curve and a gradual deterioration from the beginning to the end of the winter. The gradual change in the 'baseline' made it difficult, however, to compare the magnitude of the peaks at different times in the winter or in different winters. To overcome this difficulty, an alternative method of assessing the diary results was introduced. Each diary entry was compared with that on the preceding day, and the percentage of patients who became worse was calculated. The results are shown in Figure 5. In a period of high pollution at the beginning of December 1957, there was a large increase in the percentage worse, to 28%, and several smaller peaks in the illness curve also coincided with increases in pollution.

Temperature and humidity were also plotted but, as in the earlier study, these were not as closely related to illness as were the indices of pollution. When the 1955–56 results were re-calculated on the 'percentage worse' basis, the maximum in the illness curve was 28%, in the period of high pollution in January 1956.

As a 'control' experiment, a small group of bronchitic patients living in Crawley new town and a larger group of ex-miners working in a rural area of South Wales were studied in 1957–58. In neither case was there any uniform change in condition at any time during the winter.

The use of groups of subjects with limited mobility, whose pollution exposure could be defined more precisely than that of the general population, was also explored in 1957–58. Bronchitic patients living in Salvation Army homes responded in a uniform manner to periods of high pollution, and in 1958–59 this approach was extended to men in London prisons. These too gave a uniform response, and, among a total of 36 selected patients followed in that winter,

40

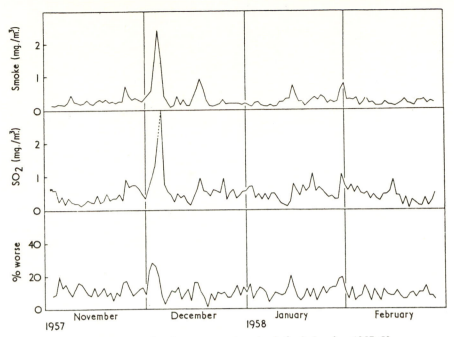

FIG. 5. *Proportion of patients 'worse than day before', London, 1957–58.*

54% became worse in the period of high pollution at the end of January 1959.

LARGE-SCALE STUDIES

The preliminary studies undertaken between 1954 and 1959 had shown that at times of high pollution changes in the condition of bronchitic patients could be demonstrated using the diary technique with small groups of about 20 carefully selected patients or with larger groups of about 200 patients selected on a more general basis. In either case there was a definite response in every episode of high pollution (when concentrations of smoke or sulphur dioxide exceeded 1,000 μg./m.^{3}), but the data were not sufficient to establish the lowest concentrations at which adverse effects occurred. A larger study was planned for the winter 1959–60, when the opportunity was taken to try a modified form of diary entry. Patients were enrolled mainly through Chest Clinics in the Greater London area, but a few were enrolled through general practitioners and prison medical officers. The criterion adopted for the selection of patients was 'those whose symptoms of chronic bronchitis, emphysema or asthma were likely to

41

be made worse by air pollution'. Since we were anxious to include a large number of patients in this study, we did not restrict the choice in any other way, and all patients who lived or worked in Greater London were accepted. In all, 1,395 patients were enrolled in the study. Each was given a serial number; those with odd numbers received a diary with instructions to enter A, B, C or D as in earlier studies and those with even numbers received one with the following instructions:

Write BETTER if your condition has been BETTER than the day before.

Write SAME if your condition has been the SAME as the day before.

Write WORSE if your condition has been WORSE than the day before.

Although in the 'ABCD' system the patients were required to compare their current with their 'usual' condition, we proposed to assess the entries by noting whether they were better, same or worse than the day before. The new system allowed the patients to do this directly, and it was more flexible than the old since they could, where appropriate, indicate a continuous deterioration day by day without limit. We did not know, however, whether the patients would find it easier or more difficult to follow, and the results from the two systems were assessed separately in the first instance. The diaries were sent out at the end of October 1959, so that most patients had them in use by early November, and new diaries were issued to run from 1 January 1960. The study was concluded at the end of March and diaries covering the whole or part of the winter were received from 1,071 patients. There appeared to be no difference in the degree of co-operation received from patients using the two systems: 539 diaries were received from those entering 'ABCD' and 532 from those entering 'better, same, worse'. Some of the patients who failed to complete or return their diaries had died or moved out of the district since their last contact with a Chest Clinic. There was no complete follow up of non-respondents.

All the results were assessed in terms of 'per cent worse', and comparison of the daily figures obtained with the two diary systems showed close agreement throughout the range: the results were therefore combined in all subsequent analyses. Entries from all returned diaries were accepted, even if they had been filled in for only a short period. Some patients showed little or no varia-

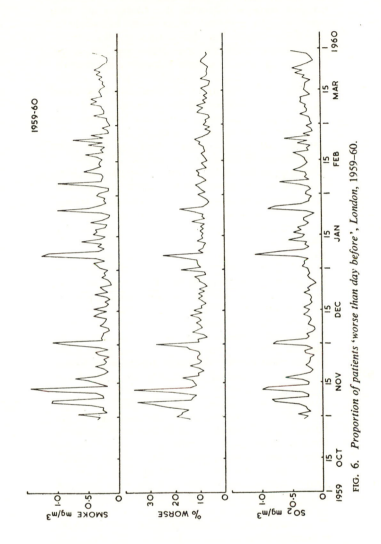

FIG. 6. *Proportion of patients 'worse than day before', London, 1959–60.*

43

tion in condition throughout the winter, but their inclusion only 'diluted' the results from other patients and did not affect the position of peaks. In Fig. 6 the mean concentrations of smoke and sulphur dioxide at seven sites in Inner London, maintained by the Greater London Council, have been used to provide indices of pollution. These sampling sites are in the Boroughs of Lambeth, Kensington, and Chelsea, Camden, Hackney, Greenwich, and Lewisham: they were originally chosen to represent conditions in the inner residential areas of London, and they have remained in operation continuously since 1957. Results from these sites were used also by Gore and Shaddick (1958) in their study of daily variations in mortality in the County of London, and by Martin and Bradley (1960) in similar studies in Greater London.

These results showed that there was a remarkably consistent response to pollution. On each occasion when the concentration of smoke or sulphur dioxide exceded 1,000 μg./m.^{3} there was a sharp increase in the percentage of patients reporting that they were 'worse' than the day before. The lowest concentration associated with any definite change was about 600 μg./m.^{3}, but there was a tendency for the association to disappear towards the end of the winter. At the beginning of the study the 'baseline' was about 12·5% worse, with peaks up to 37%, and at the end the 'baseline' was only 7·5%, with no sharp peaks, even with moderate increases in pollution. As before, we could not determine whether one pollutant was more important than another in producing the observed effects. The concentrations of smoke and sulphur dioxide followed one another very closely and either could be considered as an index of pollution: they were also very similar numerically. Graphs showing daily variations in temperature and humidity were prepared, but these showed nothing like the close association demonstrated between pollution and per cent worse.

To investigate the relative importance of smoke and sulphur dioxide, a further study was made five years later, in 1964–65, by which time the operation of the Clean Air Act had led to a substantial reduction in smoke concentrations in London, but relatively little change in sulphur dioxide. Patients were enrolled through Chest Clinics as before. The diaries were of the same type, but they were specially printed to cover the whole of the winter, from October to March. All

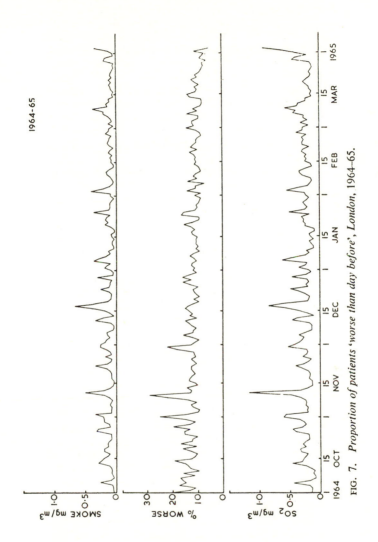

FIG. 7. *Proportion of patients 'worse than day before', London, 1964-65.*

45

patients were asked to use the 'better, same, worse' system, as this had proved to be as satisfactory as 'A,B,C,D,' and it simplified the analysis of results. Diaries were sent to 1,395 patients (the same number as in 1959–60), and in April 1965 1,037 (74%) were returned at least partially filled in. There was no need to send out new diaries half way through the winter, but letters were written to the patients then asking them to return a postcard confirming their co-operation. Entries from the diaries were punched directly on to 80 column cards (3 cards per patient). Complete tabulations were printed from the cards and daily results were worked out by computer. The diaries were sent out a few weeks earlier than in 1959–60 so that results for a full 6-month period, beginning on 1 October 1964, could be analysed.

The mean concentrations of smoke and sulphur dioxide at the seven G.L.C. sites were used as indices of pollution, as in 1959–60. As anticipated, there was less smoke than there was in 1959–60 but the average concentrations of sulphur dioxide were similar in the two winters. The diary results for 1964–65, expressed as 'per cent worse', are shown in Figure 7.

There were fewer days of high pollution in 1964–65 than in 1959–60, owing to differences in weather conditions, so that it was difficult to make any strict comparison between the results from the two studies. On the one occasion when sulphur dioxide exceeded 1,000 μg./m.^{3} there was a big increase (to 29%) in per cent worse. There was again a gradual decline in the 'baseline' from about 12·5% at the beginning to about 7·5% at the end of the period of observation. There was also some evidence that the association with pollution declined during the winter: in particular there was no response at all to a sudden increase in pollution on April 3. The concentration of sulphur dioxide then was 900 μg./m.^{3} (next to highest for the winter) and that of smoke was 300 μg./m.^{3}. This was the last day of the study. Patients were asked to send their diaries back on April 1, but about half of them were still making entries up to April 3. On another day in the middle of the winter (17 December), when the concentration of sulphur dioxide and of smoke was within the range 600 to 1,000 μg./m.^{3}, there was no response among the patients, but there were several other days with concentrations of sulphur dioxide close to 600 μg./m.^{3} when fairly well-defined peaks occurred in the 'per cent worse' curve. When the two winters were compared in

46

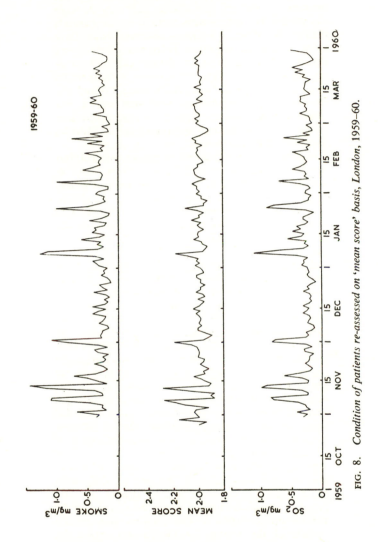

FIG. 8. *Condition of patients re-assessed on 'mean score' basis, London, 1959-60.*

terms of the association with sulphur dioxide, the general impression was of a slightly reduced and less consistent response in 1964–65 as compared with 1959–60. It was clear, however, that some association remained which had not been reduced in proportion to the reduction in smoke concentrations.

To examine the results from the two major studies on a more quantitative basis, the diary entries were reassessed on a revised scoring system. One of the problems with the original (1955–56) system, in which patients compared their daily condition with their 'usual condition', was the substantial change in baseline during the winter. The 'per cent worse' system of assessment, based on comparisons with the previous day, was devised to overcome this. It did not solve the problem completely, for the gradual increase in the illness score was replaced by a gradual decrease in 'per cent worse'. Each of these observations was of interest, for the demonstration of a deterioration in the health of bronchitic patients from November to February (Fig. 1) was in accordance with clinical impressions, and the reduction in 'per cent worse' (Figs 6 and 7) indicated that the condition of the patients tended to become more stable during the course of the winter. This was not due to any selective withdrawal of the more variable patients from the study. The percentage reported 'better' also declined gradually through each winter and it seemed likely that the patients were either losing interest or that when they reached their lowest winter level they no longer responded to changes in the environment. By assigning 'scores' to the individual entries (better = 1, same = 2, worse = 3) and calculating mean scores day by day it was found that the trends in 'per cent worse' and 'per cent better' cancelled out, yielding graphs (Figs 8 and 9) that had essentially level baselines.

The mean values of all the quantities plotted in Figs 8 and 9 are shown in Table I, together with the corresponding standard deviations, and correlation coefficients between mean score and the indices of pollution.

In each winter the overall mean score was close to 2, as expected, though there was a small decline from November to March. The standard deviation of the mean score was greatest in November, and (as shown in Table II) there were then more days of high pollution than in other months. Throughout the winter of 1964–65 concentrations of smoke

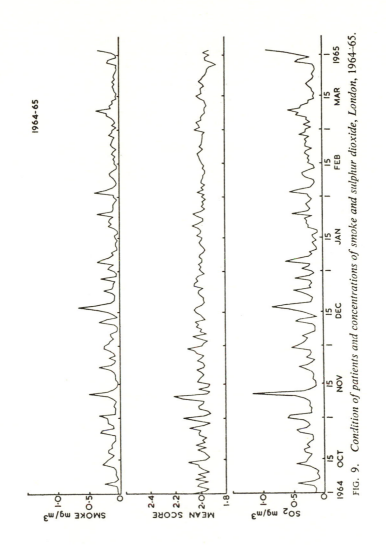

1964-65

SMOKE mg/m3   1·0   0·5   0

MEAN SCORE   2·4   2·2   2·0   1·8

SO2 mg/m3   1·0   0·5   0

1964   OCT 15   NOV 15   DEC 15   JAN 15   FEB 15   MAR 15   1965

FIG. 9. *Condition of patients and concentrations of smoke and sulphur dioxide, London, 1964-65.*

TABLE I

ANALYSIS OF DIARY RESULTS, WINTERS 1959-60 AND 1964-65

	No. of Days	Mean Score		Smoke (μg./m.³)		SO$_2$ (μg./m.²)		Corr. Coeff.	
		Mean	S.D.	Mean	S.D.	Mean	S.D.	Score/smoke	Score/SO$_2$
1959–60									
October	6	2·03	0·07	357	175	274	99	0·44	0·60
November	28	2·00	0·10	435	340	349	240	0·36	0·42¹
December	35	1·99	0·05	294	185	248	143	0·36¹	0·53¹
January	28	2·00	0·06	393	284	365	235	0·61¹	0·62¹
February	28	1·99	0·03	340	183	285	129	0·37	0·51¹
March	31	2·00	0·02	264	83	254	89	−0·39¹	−0·29
Whole winter	156	2·00	0·06	342	232	296	177	0·38¹	0·44¹
1964–65									
October	35	2·00	0·06	144	85	262	129	0·17	0·33
November	28	2·01	0·06	131	99	275	199	0·10	0·14
December	35	2·00	0·03	158	127	292	157	0·24	0·29
January	28	1·99	0·03	117	87	228	121	0·36	0·18
February	28	1·99	0·02	109	70	252	88	0·24	0·23
March	33	1·97	0·03	110	80	272	173	−0·17	0·25
Whole winter	187	1·99	0·04	129	95	264	149	0·16¹	0·13

The results were analysed by four or five week periods rather than calendar months.
The first period in 1959–60 and the last in each winter were incomplete.
¹ Significant at 5% level.

were well below those of 1959–60: the mean concentration during the period of the later study was only 38% of that five years earlier. There was, however, little change in the mean concentration of sulphur dioxide. Bearing in mind the earlier reservations made about the complex time relationships, the correlation coefficients between mean score and pollution were quite high, particularly in the winter of 1959–60. Those between mean score and sulphur dioxide were significant for each of the four months November 1959 to February 1960. The coefficients were much lower in 1964–65 than they were in 1959–60, and none of those for monthly periods reached the 5% level of significance.

An attempt to compare the response from month to month and from one winter to the other has been made in Table II. For this purpose days on which the mean score was more than twice the (whole winter) standard deviation above the mean were tabulated with the corresponding concentrations of smoke and sulphur dioxide. This procedure selected the main sets of coincident peaks seen in Figs 8 and 9; in general, the mean score and pollution peaks coincided exactly, but some were out of phase by one day. Peaks in October, before all the patients had been enrolled, were excluded.

Other days on which the concentration of smoke or sulphur dioxide exceeded 500 μg./m.^{3} have been included in Table II, together with corresponding 'peaks' in mean score. The response to pollution clearly declined during the course of each winter, and it would therefore be unrealistic to compare the results from an episode near the beginning of one winter with those near the end of the other. For this reason the episodes have been arranged in date order in Table II, placing those occurring at roughly the same time in each winter side by side to facilitate comparison. The deviations observed in 1964–65 were, in general, lower than those seen in 1959–60, even when the concentrations of sulphur dioxide were similar (as in the second pair). There were not enough pairs of episodes to determine an exact relationship between the 'response' of patients and the concentrations of smoke and sulphur dioxide, but the main findings can be summarized as follows:

1. Patients are most sensitive to changes in pollution at the beginning of each winter, i.e., in November.

2. The minimum pollution leading to any significant response is about 500 μg./m.^{3} of

51

TABLE II

PEAK VALUES IN DIARY STUDIES, 1959–60 AND 1964–65

1959–60				1964–65			
Date	Smoke (μg./m.³)	SO$_2$(μg./m.³)	Deviation	Date	Smoke (μg./m.³)	SO$_2$(μg./m.³)	Deviation
7 November	1,095	786	0·28*	1 November	209	599	0·14*
12 November	1,162	990	0·29*	10 November	499	1,160	0·22*
17 November	708	612	0·04				
1 December	810	809	0·20*	30 November	262	524	0·11*
				16 December	660	838	0·06
7 January	1,256	1,095	0·19*	4 January	353	614	0·02
13 January	602	546	0·04				
26 January	999	875	0·11	3 February	402	543	0·03
5 February	977	675	0·05				
24 February	736	595	0·05	9 March	369	554	–0·03
				2 April	345	912	–0·06

In this table, occasions on which pollution and/or mean score was high have been isolated, using the following criteria: smoke or SO$_2$ greater than 500 μg/m.³; mean score — deviation from winter mean greater than twice the standard deviation (indicated by*). Dates refer to the pollution peaks: the dates of the mean score peaks differed from these by one day in a few cases.

sulphur dioxide together with about 250 μg./m.^{3} of smoke (each representing the average, over 24 hours, at a group of sites in Inner London). It is important to recognize, however, that there is no evidence that either of these pollutants would, by itself, produce the same response.

3. The type of pollution found in London now, with much less smoke, and fewer days of high pollution of any kind, has led to a reduced response among the patients.

When individual diary entries were examined, it was found that many of the patients did not respond at all to episodes of high pollution, and others responded to just one or two of the episodes shown in Table II. One possible explanation was that patients were affected by the first period of high pollution that they encountered in each winter, and thereafter became relatively insensitive, due either to restriction of activities once they became ill or to protection afforded by therapy or by increased production of mucus. Some may have escaped exposure to the earlier episodes, leaving a diminishing 'pool' of susceptibles to respond to later ones. An attempt was made to separate from the records patients who appeared to respond to several episodes of high pollution. Those who became worse on at least one-third of the occasions when the concentration of smoke or sulphur dioxide exceeded 500 μg./m.^{3} and who did not vary much at other times were identified: 307 such patients were selected in 1959–60, and 87 in 1964–65. Such combinations of results could have occurred by chance, but it was of interest to consider whether these patients were particularly sensitive to changes in pollution. There was nothing obviously different about them in respect of sex, age, or area of residence.

When the final study in the present series was planned for the winter 1967–68, these selected patients were used again to determine whether they might in fact be useful 'monitors' of the effects of pollution. Of the 87 patients selected, 50 were able and willing to participate again. To reduce the risk of patients filling in their diaries for long periods from memory (and evidence that some had done this had been seen in the earlier studies), weekly postcards were issued, drawn up for daily entries in the same way as the diaries, and stamped ready for return. The 'lapse' rate was reduced in this way, for enquiries were made immediately if postcards failed to arrive when expected. Results were graphed week by week, though it was disappointing to note that there

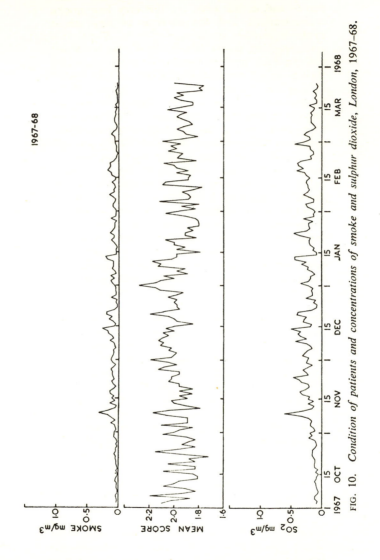

FIG. 10. *Condition of patients and concentrations of smoke and sulphur dioxide, London, 1967–68.*

54

were hardly any periods of high pollution to test responses. Pollution was measured at the same sites as before, and the results are shown in Figure 10.

The outstanding differences between these curves and earlier ones are the very low smoke concentrations and the absence of any major peaks in smoke or sulphur dioxide. There was only one day in the whole of the winter when the concentration of sulphur dioxide exceeded 500 μg./m.^{3} (9 November: SO_2 605 μg./m.^{3}, smoke 317 μg./m.^{3}). The peak in the mean score curve on that day represented a deviation that was not quite twice the standard deviation, and there were other peaks of this magnitude that did not correspond with days of high pollution. There was still some overall association with pollution, and in Table III the correlation coefficients are compared with those obtained from the same (selected) group of patients in 1964–65.

The 1964–65 coefficients had been enhanced by the selection procedure, but, as in the whole population for that year (Table I), the correlation with smoke was a little higher than that with SO_2. The relationship with several other variables was also studied; in particular, the correlation with daily

TABLE III

SUMMARY OF RESULTS, SELECTED PATIENTS, 1964–65 AND 1967–68

		1964–65	1967–68
Mean score	Mean	1·98	1·96
	S.D.	0·10	0·11
Smoke (μg./m.^{3})	Mean	129	68
	S.D.	95	48
SO_2 (μg./m.^{3})	Mean	264	204
	S.D.	149	100
H_2SO_4 (μg./m.^{3})	Mean	7·3	6·3
	S.D.	4·8	4·0
Temp. (°C.)	Mean	6·4	6·3
	S.D.	4·1	4·3
Corr. coeff., mean score and	Smoke	0·39[1]	0·31[1]
	SO_2	0·30[1]	0·28[1]
	H_2SO_4	0·51[1]	0·26[1]
	Temp.	−0·24[1]	−0·17[1]

These results are for the whole winter period, October to March.
[1] Significant at 5% level.

measurements of particulate sulphuric acid, made at our own laboratory, was found to be relatively high. There was also a negative correlation with temperature (as measured in Central London at 9 a.m. each day). For 1967–68, all these correlation coefficients were lower but still significant. This time the correlation with sulphuric acid was lower than that with other pollutants. It seemed likely that the patients selected were particularly sensitive to pollution, for from past experience no correlation would have been expected with the very low levels of pollution encountered by such a small group.

DISCUSSION

One of the purposes of the present paper was to demonstrate the way in which a simple and inexpensive method of enquiry had led to useful indications of the effects of environment on health. This 'diary' technique is applicable only to the study of patients with diseases in which frequent changes in condition may occur. Apart from its value in studies of chronic bronchitis in relation to air pollution, it has been used to study variations of the severity of rheumatoid arthritis with changes in weather. A similar approach has also been used to assess the results of treatment among asthmatic patients (McAllen, Heaf, and McInroy, 1967; British Tuberculosis Association, 1968). As it is essential to maintain the patients' interest and co-operation, efforts have been made to find out which method of recording (on printed sheets, postcards, or pocket diaries) the patients preferred. Those who had tried more than one system favoured small pocket diaries that they could carry around with them. The use of postcards returned once a week helped us but was not popular among the patients. The diaries that were specially printed[1] to cover a whole winter season were the most convenient.

'Wandering baselines' were minimized by the use of questions relating to a change in condition compared with the day before. Only one simple question each day on general condition should be asked, since the inclusion of additional questions only tended to confuse the patient. The presence of a fairly large proportion of patients who do not respond at all to changes in the environment 'dilutes' the associations but does not cancel out the response of the others. Although we were able to select 'sensitive' patients from one study (1964–65) for use in another (1967–68), this procedure is of limited value, since their co-operation falls off if the same patients are asked to continue for more than one season; furthermore, after an interval of several years many have moved, entered hospital or died.

The results reported here demonstrate convincingly that pollution rather than adverse weather is associated with exacerbations of chronic bronchitis. So far we have not been able to show which pollutant or combination of pollutants is responsible for the associated exacerbations, and the concentrations of smoke and sulphur dioxide used can only be regarded as indices of the active agents. Attempts to assess the relative importance of smoke and sulphur dioxide have been frustrated by an absence of periods of high pollution of any kind in recent years. This improvement is in itself important; the loss of the 'blanket' of smoke over London may have reduced the tendency for temperature inversions to persist, thereby reducing the risk of high concentrations of any pollutant being reached. Comparisons between the two main studies, done five years apart, show a reduced overall response to pollution during the later winter, when there was less smoke and few periods of high pollution. The interpretation of the findings is difficult, since the susceptibility of patients to the effects of pollution may have changed as a result of improved drug therapy. Studies on mortality and hospital admissions in London also show that there is no longer a close relationship with pollution during the winter months, as there was some 10 years ago, and it is clear that the net result of the changes in pollution has been beneficial (Waller, Lawther, and Martin, 1969). The concentration of smoke in London is still declining year by year, and in future studies it may be possible to observe the effects of what sulphur dioxide remains when concentrations of smoke have become very low in comparison with those of 10 to 15 years ago.

The limitations of this technique must be recognized. Some authors have used results from epidemiological studies as guides for the establishment of 'air quality criteria' in respect of sulphur dioxide (U.S. Dept. of Health, Education and Welfare, 1969). This can be misleading, for it is only possible to assess the response of patients to the general mixture of pollutants in the air. In London, even with the disappearance of much of the black smoke, there are still other pollutants present apart from sulphur dioxide. Among these, sulphuric acid may be of special interest as a respiratory irritant, and we now have daily measurements of this pollutant to link with epidemiological data. It should also be stressed that the absolute concentrations of pollutants, as measured in our surveys, serve only as a guide to

[1] These diaries were supplied by J. M. Tatler. Printer, Abbey Street Works, Derby

the exposure of patients. In particular, all the measurements were made out of doors, whereas many of the patients would have spent much time indoors. The concentrations found indoors depend on the type of heating system, the degree of ventilation and other characteristics of the house, but in general the concentration of smoke is fairly close to that outdoors, whilst the concentration of sulphur dioxide is appreciably lower indoors. Biersteker, de Graaf, and Nass (1965) reported concentrations of sulphur dioxide in homes in Rotterdam that were on average only 20% of those out of doors, but they found one example of a house where indoor concentrations were consistently above those outside. This can happen when faulty flues or heating appliances allow fumes to escape into the room, and Biersteker et al. (1965) consider that this may be of some importance in episodes of high pollution, when there is little wind to induce draughts in chimneys.

One further reservation is that the measurements quoted in the present paper relate only to 24-hour average concentrations of sulphur dioxide and other pollutants. The evidence that we have from experimental work on normal subjects suggests that the effects of inhaling prepared mixtures of pollutants are of rapid onset, and peak concentrations encountered during the day may therefore be more relevant than 24-hour averages. The effects that we have reported cannot be considered as the result of 24-hour exposures to at least 500 μg./m.^{3} of sulphur dioxide together with 250 μg./m.^{3} of smoke: they are more likely to reflect the effects of brief exposures to the maximum concentrations occurring during the day, and these may be several times the 24-hour averages (Waller and Commins, 1966).

In this series we have not tried to differentiate between various respiratory symptoms, nor have we attempted to make any objective measurements of changes in lung function. Daily observations on ventilatory capacity in normal subjects and in a small number of bronchitic patients have been made in a separate study (Lawther, Brooks, and Waller, to be published), and in some individuals there is evidence of an effect of pollution. It is not, however, practicable to make daily measurements, even with the relatively simple peak flow meter, on large groups of subjects, and in a third study in our 'five-yearly' series which is now in progress (1969–70) we are again confining ourselves to the use of the simple diary technique described above.

In the course of the studies reported here we have enjoyed the enthusiastic co-operation of the physicians and staffs of more than 60 Chest Clinics and Hospitals in the London area, plus others in Sheffield, Manchester, Birmingham, and Wolverhampton. We are also indebted to general practitioners in Greater London and in Crawley who helped in the selection of patients, to Officers of the Salvation Army (London Division), members of the Prison Commission Medical Service, the Manager and staff of the Austin Junior Car Factory, Hengoed, and to members of the staff of the Ferguson Radio Corporation factory at Enfield.

Many of our colleagues in the Air Pollution Unit have assisted in the organization of the surveys, and the control study in South Wales was arranged by the M.R.C. Pneumoconiosis Unit. Access to records of pollution measurements was freely granted by the Greater London Council and the Warren Spring Laboratory, Stevenage. Mr. H. Kasap, of St. Thomas's Hospital, kindly prepared a program for the analysis of the results, and this was run on the IBM 7094 computer at Imperial College.

Finally, we are grateful to the subjects (over 3,000 in all), most of whom diligently filled in their diaries day by day, offering in addition many useful comments on their condition and on the conduct of the survey.

REFERENCES

Biersteker, K., de Graaf, H., and Nass, Ch. A. G. (1965). Indoor air pollution in Rotterdam homes. *Int. J. Air Wat. Pollut.*, 9, 343.

Bradley, W. H., Logan, W. P. D., and Martin, A. E. (1958). The London fog of Dec. 2-5, 1957. *Mth. Bull. Minist. Hlth Lab. Serv.*, 17, 156.

British Tuberculosis Association (1968). Hypnosis for asthma—a controlled trial. *Brit. med. J.*, 4, 71.

Clifton, M. (1967). Pollution data for health studies. In *Trans. Int. Chest Heart Conf., Eastbourne, 1967*, p. 143. Chest and Heart Association, London.

Gore, A. T., and Shaddick, C. W. (1958). Atmospheric pollution and mortality in the County of London. *Brit. J. prev. soc. Med.*, 12, 104.

McAllen, M. K., Heaf, P. J. D., and McInroy, P. (1967). Depot grass-pollen injections in asthma: effect of repeated treatment on clinical response and measured bronchial sensitivity. *Brit. med. J.*, 1, 22.

Martin, A. E. (1961). Epidemiological studies of atmospheric pollution. *Mth. Bull. Minist. Hlth Lab. Serv.*, 20, 42.

—— and Bradley, W. H. (1960). Mortality, fog and atmospheric pollution. *Mth. Bull. Minist. Hlth Lab. Serv.*, 19, 56.

Ministry of Health (1954). *Mortality and Morbidity during the London Fog of December 1952. Rep. publ. Hlth and med. Subjects* no. 95. H.M.S.O., London.

U.S. Dept. of Health, Education and Welfare (1969). Air Quality Criteria for Sulphur Dioxide. National Air Pollution Control Administration, Washington, D.C.

Waller, R. E., and Commins, B. T. (1966). Episodes of high pollution in London, 1952-1966. In *International Clean Air Congress, London, 1966. Proceedings*, Part 1, p. 228. National Society for Clean Air, London.

—— and Lawther, P. J. (1955). Some observations on London fog. *Brit. med. J.*, 2, 1356.

—— —— (1957). Further observations on London fog. *Brit. med. J.*, 2, 1473.

—— —— and Martin, A. E. (1969). Clean air and health in London. Clean Air Conf., Eastbourne, 1969. Part I. Pre-prints of Papers, p. 71. Nat. Soc. Clean Air, London.

Warren Spring Laboratory (1966). National Survey of Smoke and Sulphur Dioxide. Instruction Manual. Warren Spring Laboratory, Stevenage.

STATISTICAL NOTE ON ASSOCIATION
OF AIR POLLUTION AND LIVER CIRRHOSIS

S. C. Morris, MS
M. A. Shapiro, MEng

To the Editor.—A statistical analysis was performed on data reported by Winkelstein and Gay in the Archives (22:174, 1971). The data relate suspended particulate air pollution, economic level, and death rate from cirrhosis of the liver for the Buffalo area during the period, 1959 to 1961. Examination of the results of the analysis yields some interesting insight into the association of air pollution and cirrhosis.

The Analysis.—The problem of analysis presented by the data is one that does not receive sufficient emphasis in texts or courses in biostatistics, yet it is an important and common one in environmental epidemiology. Winkelstein and Gay's Table 3 and 4 present data that at first seem to be suited to analysis of variance as a two-way classification. Two problems make this type of analysis impractical. First, there is the fact that the number of people at risk is different in each cell. The cell population then becomes an additional variable requiring consideration. (Cell populations can be calculated, since the total deaths and the death rate within each cell are reported in the Winkelstein and Gay article.) The second problem is that of the four missing cells. This not only complicates the analysis but seriously detracts from the validity of the results.

A different approach to analysis was,

therefore, selected. The null hypothesis that mortality rates from cirrhosis were independent of air pollution was tested. The expected number of deaths was determined by applying the death rate for each economic level of the population at each air pollution level within that economic level, and a goodness-of-fit chi square value was calculated. This analysis is shown in Table 1 and 2.

Results.—The data for white women 50 years old and older do not show a significant correlation at the .05 level between mortality from cirrhosis and air pollution, either for the entire population or within any economic level. Analysis of the data for white men 50 years old and older results in an overall chi-square value indicating an extremely significant relationship. Further inspection, though, shows a relationship within economic blocks significant at the .05 level only for the two low economic levels. Inspection of the individual cells reveals that half of the overall chi-square value is contributed by two cells (column 1, line 4, and column 2, line 1) which comprise only 8% of the total population cells. Even with these two cells rejected, however, the relation between air pollution and cirrhosis mortality would remain significant at the .05 level both within the two lower economic levels and in the overall population.

Comment.—The lack of a significant relation within the upper economic levels might suggest the possibility of a causal airborne toxic agent that is present at greater levels in the polluted low-economic-level areas, rather than being distributed in proportion to total suspended particulate matter in the air. This could result from the nature of the source of the pollutant in general or from particular characteristics of the Buffalo area. If this were the case, though, a significant correlation would be expected for both men and women in the lower economic level areas. Since there was not a significant correlation for women in the low economic levels, more likely explanations would seem to be a

Table 1.—Deaths From Liver Cirrhosis For White Women 50 Years Old and Older in Buffalo and Environs, 1959-1961*

Economic Level	No.	Air Pollution Level				Statistical Analysis†
		1	2	3	4	
1	Observed	...	0	8	4	$\chi^2 = 2.59$
	Expected	...	1.3	8.5	2.3	df = 2
	(O-E)²/E	...	1.30	0.03	1.26	
2	Observed	2	9	9	6	$\chi^2 = 4.97$
	Expected	4.4	11.4	7.2	3.1	df = 3
	(O-E)²/E	1.31	0.50	0.45	2.71	
3	Observed	...	6	4	3	$\chi^2 = 3.27$
	Expected	...	8.1	3.7	1.2	df = 2
	(O-E)²/E	...	0.54	0.03	2.70	
4	Observed	8	8	3	...	$\chi^2 = 0.46$
	Expected	6.8	9.5	2.8	...	df = 2
	(O-E)²/E	0.21	0.24	0.01	...	
5	Observed	3	5	0	...	$\chi^2 = 1.05$
	Expected	3.5	3.8	0.6	...	df = 2
	(O-E)²/E	0.07	0.38	0.60	...	

$$\chi^2 = \Sigma \cdot \left[\Sigma \frac{(O\text{-}E)^2}{E} \right] = 12.34$$
$$df = 15$$

* Expected death rates computed on basis of null hypothesis that mortality rates are independent of air pollution level.
† df, Degrees of freedom.

Table 2.—Deaths From Liver Cirrhosis for White Men 50 Years Old and Older in Buffalo and Environs, 1959-1961*

Economic Level	No.	Air Pollution Level				Statistical Analysis†
		1	2	3	4	
1	Observed	...	3	28	28	$\chi^2 = 16.55$
	Expected	...	4.3	40.0	14.5	$df = 2$
	$(O-E)^2/E$	...	0.34	3.60	12.61	
2	Observed	4	32	38	21	$\chi^2 = 19.01$
	Expected	14.6	39.5	27.2	12.6	$df = 3$
	$(O-E)^2/E$	7.74	1.42	0.02	5.61	
3	Observed	...	14	13	4	$\chi^2 = 2.66$
	Expected	...	18.3	9.4	3.1	$df = 2$
	$(O-E)^2/E$	...	1.01	1.39	0.26	
4	Observed	9	10	8	...	$\chi^2 = 4.55$
	Expected	10.4	12.9	4.1	...	$df = 2$
	$(O-E)^2/E$	0.19	0.65	3.71	...	
5	Observed	6	4	2	...	$\chi^2 = 5.09$
	Expected	5.6	5.8	0.5	...	$df = 2$
	$(O-E)^2/E$	0.03	0.56	4.50	...	

$$\chi^2 = \Sigma \left[\frac{(O-E)^2}{E} \right] = 47.90$$
$$df = 15$$

* Expected death rates computed on basis of null hypothesis that mortality rates are independent of air pollution level.
† df, Degrees of freedom.

synergistic effect of urban air pollution and one or more other agents to which low-economic-level men are selectively exposed or a noncausal association between low-income men and cirrhosis mortality.

Conclusion.—As a result of a statistical analysis, the data presented by Winkelstein and Gay can be more meaningfully interpreted. Since a significant relationship exists between air pollution levels and cirrhosis mortality only for men in the lower economic levels, it seems reasonable to conclude that air pollution is not a direct cause of cirrhosis, but may act in a synergistic manner with other causal agents.

In Vivo and In Vitro

Ciliotoxic Effects of Tobacco Smoke

Tore Dalhamn, MD

Rabbit trachea in vitro and cat trachea in vivo has been exposed to 1-ml and 10-ml tobacco smoke. The results show that 1-ml exposure required 73 puffs and 71 puffs to produce ciliostasis in rabbit trachea and cat trachea, respectively. In the experiments with 10-ml puffs, 37 puffs and 35 puffs were required to produce ciliostasis. The implications are that the technique of preparing the specimens may be essentially simplified and that fewer animals should be required with use of the vitro technique.

THE ACUTE effects of tobacco smoke on the respiratory mucosa have been studied with many methods in many animal species. The importance of various experimental factors such as the technique of exposure to the smoke, the species of the test animals, and in vivo or in vitro observations was pointed out in a recent paper on "the experimental requirements for the toxicological evaluation of tobacco smoke in the respiratory system."[1] The available literature contains no direct experimental comparisons of these variables, and extrapolation of findings in animals to man demands great caution. Comparisons of ciliotoxicity in different animal species and between results in vivo and in vitro, therefore, seem to offer a promising field for research. The following is an account of a pilot investigation in this field.

Materials and Methods

Observation Technique.—The in vivo technique was identical with that earlier described.[2] In brief, it implies that the animals were anesthetized with pentobarbital sodium (Nembutal), after which the trachea was exposed and incised. The tracheostomy was connected to a specially constructed rubber bellows and to a direct-light microscope. All connections were airtight. The ciliary beating in the trachea was observed through the microscope.

In the in vitro observations, a small piece of tracheal tissue was placed in a moist, warm chamber, and the ciliary activity was studied by direct-light microscopy. These experimental arrangements have also been previously described.[3]

Exposure Technique.—From the outset, it was deemed essential that in in vivo exposure to tobacco smoke the animals should inhale the smoke through the mouth and not through the nose since undesirable absorption phenomena might occur in nose breathing. The trachea was, therefore, cut just below the larynx and was fitted with an artificial "mouth," ie, a space volume approximating to that of the animal's oral cavity. This mouth was moistened and warmed to conform with natural conditions. Respiration took place through a Sterling pump.

The tobacco smoke was conducted into a special opening in the artificial mouth. From a 35-ml "standard puff" in a syringe, fixed volumes were introduced into the mouth during the inspiratory phase of respiration. The rate of "smoking" was one puff per minute, and the cigarettes were smoked to a butt length of 28 mm.

The exposure was continued until ciliostasis occurred. The number of puffs preceding ciliostasis was recorded.[4]

In in vitro exposure to tobacco smoke, the volume of the puffs, and the intervals between the puffs were the same as in the experiments on living animals. Thus, the desired smoke volume was taken from a 35-ml "standard puff" and was introduced in the moist chamber (volume of the chamber, approximately 40 ml) once per minute. The butt length of the cigarettes was 28 mm. The number of puffs preceding ciliostasis was calculated.

Animal Material.—Cats were used in the in

vivo experiments. Rabbit tracheas were studied in vitro, about six pieces being cut from each trachea. Two smoke volumes were tested in both types of experiment, namely 1 ml and 10 ml. In the experiments using 1 ml, 20 live cats

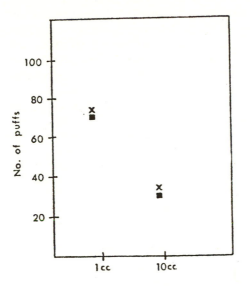

Ciliotoxicity of tobacco smoke (1 and 10 cc) on trachea of rabbits in vitro (solid square) and cats in vivo (x).

and 40 pieces of rabbit trachea were studied. The 10-ml experiments were made on 12 cats and 20 pieces of rabbit trachea.

In order to ensure maximum standardization of the experimental material, all the prepared cat tracheas and one specimen from each of the rabbit tracheas were first exposed to air or placed in the exposure chamber for one hour. If at the end of that time the ciliary activity appeared to be unchanged, the exposure to tobacco smoke was begun.

Results and Comment

In the 20 live cats which were exposed to 1-ml puffs of smoke, the mean number of puffs required to produce ciliostasis was 72.8 ± 27.1. The corresponding mean in the 40 pieces of rabbit trachea was 70.7 ± 26.9

69

(Figure). The difference between these means is not statistically significant.

In the experiments with 10-ml puffs on 12 live cats, ciliostasis occurred after on average 36.9 ± 11.2 puffs. When 10-ml puffs were blown over 20 pieces of excised rabbit trachea, a mean of 34.8 ± 11.2 puffs preceded ciliostasis (Figure). This difference likewise lacks statistical significance. Between the mean figures in the experiments with 1-ml puffs and in those with 10-ml puffs, however, the difference is clearly significant.

The reported comparisons appeared to be interesting from at least two aspects. Whether the puff size was 1 ml or 10 ml, the number of puffs preceding ciliostasis was similar in vivo and in vitro and also in rabbit and in cat trachea. In regard to ciliotoxicity, therefore, it would seem to be a matter of indifference whether in vivo or in vitro technique is used. The implications are that the technique of preparing the specimens may be essentially simplified and that fewer animals should be required since, for instance, six or more pieces may be cut from the same rabbit trachea.

Comparisons between results in large numbers of live animals and in excised tissue from the same species would undoubtedly help to standardize experiments designed to demonstrate ciliotoxic effect. The present experiments, however, were too limited to permit general conclusions.

The demonstrated differences in ciliotoxic action obtained with 1-ml and 10-ml puffs of tobacco smoke agree well with the dose response described in an earlier paper.[5]

This study was aided by a grant from the American Medical Association Education and Research Foundation.

References

1. Dalhamn T, Rylander R: Experimental requirements for the toxicological evaluation of tobacco smoke in the respiratory system. *Scand J Resp Dis* 50:273-280, 1969.
2. Dalhamn T: The determination in vivo of the rate of ciliary beat in the trachea. *Acta Physiol. Scand* 49:242-250, 1960.

3. Dalhamn T, Lagerstedt B: Ciliostatic effect of phenol and resorcinol. *Arch Otolaryng* 84:325-328, 1966.

4. Dalhamn T, Rylander R: Cigarette smoke and ciliastasis: Effect of varying composition of smoke. *Arch Environ Health* 13:47-50, 1966.

5. Dalhamn T: *Effect of Different Doses of Tobacco Smoke on Ciliary Activity in Cat: Variations in Amount of Tobacco Smoke, Interval Between Cigarettes, Content of Tar, Nicotine and Phenol.* National Cancer Institute monograph, No. 28, Bethesda, Md, 1968, pp 29-87.

Transformation of Rat and Hamster Embryo Cells by Extracts of City Smog

AARON E. FREEMAN , PAUL J. PRICE , ROBERT J. BRYAN , ROBERT J. GORDON , RAYMOND V. GILDEN , GARY J. KELLOFF , AND ROBERT J. HUEBNER

We have reported that a chemical carcinogen, diethylnitrosamine, induced morphological transformation of rat cell cultures chronically infected with murine leukemia virus (MuLV), whereas the virus alone, or the chemical alone, had no effect under our experimental conditions (1). Subsequently, it has been shown that the same patterns of sensitivity to transformation occurred upon exposure of the cultures to methylcholanthrene (2) or dimethylbenzanthracene (3). In our current project, we have utilized both rat and hamster embryo cell cultures chronically infected with leukemia viruses to study certain factors of the environment. This paper reports the morphological transformation of rat and hamster embryo cells by extracts of city smog.

MATERIALS AND METHODS

Cell lines

F111. A pool of normal Fischer rat-embryo cells was carried serially and stored at various subculture levels in the vapor phase of liquid nitrogen. The cultures were shown to be free of pleuropneumonia-like organisms (PPLO) and negative by the mouse-antibody production test for H-1, K virus, Kilham rat virus, lymphocytic choriomeningitis, minute virus of mice, mouse adenovirus, mouse hepatitis, polyoma, reovirus, Sendai, simian virus 5, and Theiler's GDVII. The cultures were also shown to be free of group-specific MuLV antigens by complement-fixation (CF) (4). The F111 cell line at subculture 13 was used at the beginning of these experiments.

F115. The F111 was infected as a primary culture with Rauscher leukemia virus (RLV). This cell line (F115) was also shown to be free of PPLO and all the previously listed viruses except RLV. The cell cultures receiving experimental treatment were also at subculture 13. Detailed characteristics of the F111 and F115 cell lines have been published previously (1).

F839. This serial cell line, derived from Syrian hamster embryo* (Lakeview), was obtained from Dr. Gary J. Kelloff of the National Cancer Institute (5). The cell line was free of PPLO and the viruses outlined above. The F839 cell cultures receiving experimental treatment were at subculture 6.

F840. This was a sister cell line of F839, but chronically infected with a hamster C-type RNA virus (HaLV). It was free of PPLO and the viruses outlined above. The F840 cell cultures receiving experimental treatment were at subculture 6.

Virus stocks

The RLV (6) used to infect F115 was obtained from Dr. Janet W. Hartley of the National Institute of Allergy and Infectious Diseases. It had been passaged 22 times in NIH Swiss-mouse embryo cultures.

The HaLV virus used for infecting the F840 cell line was isolated by Kelloff *et al.* (5) as a hamster-tropic nonfocus-forming helper virus from a hamster sarcoma originally induced by the Gross pseudotype of Moloney sarcoma.

Benzo(*a*)pyrene (BZP)

This was obtained from Eastman Organic Chemicals and recrystallized from *n*-hexane. From a stock solution made in chloroform, one 50-μl portion containing 50 μg was put in a small vial and the chloroform was evaporated with a stream of nitrogen to leave 50 μg of dry BZP. This was dissolved in

---

Abbreviations: BZP, benzpyrene; CF, complement-fixation. Viral strains: MuLV, murine leukemia virus; RLV, Rauscher leukemia virus; HaLV, hamster C-type RNA virus.

---

* This is the London School of Hygiene (LSH) strain.

73

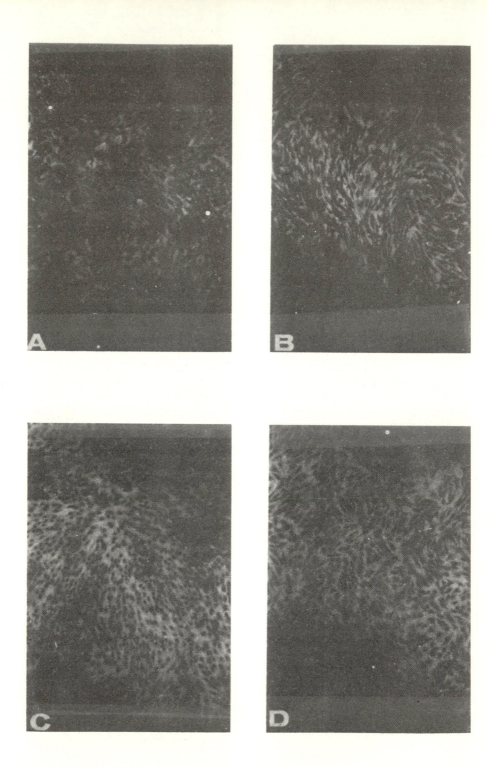

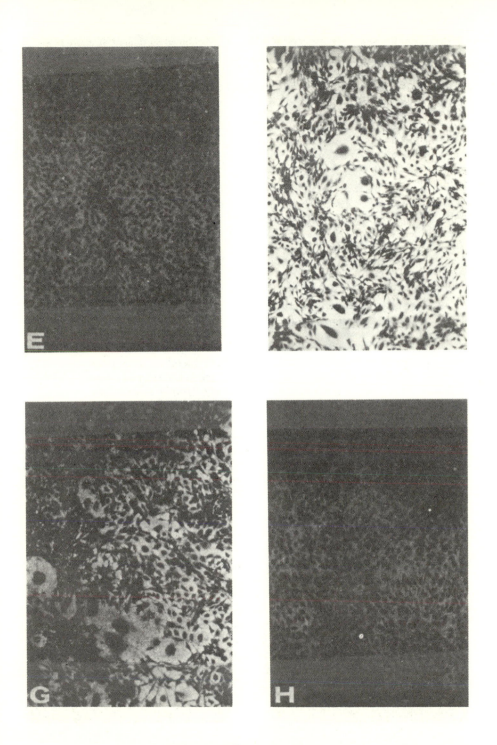

Fig. 1.

1 ml of acetone to yield a stock solution of 50 μg/ml, which was stored in the dark at 4°C.

Organic fraction of airborne particulate matter

The sample was collected on a pleated fiberglass filter from August 11 to October 6, 1969, at a central location in one of the major metropolitan areas of the United States. The total air volume sampled was 1.1×10^7 m³. A portion of the filter was extracted with benzene for 24 hr and the solvent was evaporated to a constant residual weight. The residue was equal to 16.2 μg/m³ of air. Assay by ultraviolet spectrometry showed that each gram of residue contained approximately 23 μg of BZP. The residue was dissolved in acetone to yield a stock solution of 100 μg units/ml (see below).

Experimental procedure

Replicate cultures of each of the cell lines, at the indicated subcultures, were planted at 1×10^5 cells/ml. After 3 days, when the bottles were 50–80% confluent, they were treated in triplicate with 20 ml of experimental media containing dilutions of BZP, or equivalent concentrations of BZP in smog residue, which will be referred to as μg units. A μg unit of smog contains approximately 43,000 μg of residue of which 1 μg is benzpyrene. Appropriate controls were fed equivalent concentrations of acetone in Eagle's Minimum Essential Medium with 10% fetal calf serum (FBS10). The cultures were refed with experimental media after 3 days. After a total exposure to the chemical of 6 days, the experimental media were removed permanently, and the cultures were fed with FBS10. After 24 hr the cultures were subdivided 1:2. Further subdivisions of the cultures were made as they became confluent. References to subculture numbers later in this paper indicate the number of population doublings after exposure to experimental treatment.

Tests for transplantability

The rat and hamster embryo cells (transformed and controls) were inoculated subcutaneously into newborn inbred F344/f rats and newborn Syrian hamsters (Lakeview) at a concentration of 1×10^6 cells in 0.05 ml.

RESULTS

Effects on rat embryo cultures

The RLV-free rat embryo cultures (F111) were not transformed by doses of smog up to 0.007 μg units/ml nor by the BZP control at doses up to 1.0 μg/ml. Higher dosages of either agent were lethal. In contrast, the RLV carrier cultures (F115) were transformed by both smog and BZP. Transformation was induced by as little as 0.0007 μg units/ml of smog after 2–5 subcultures under experimental conditions and 0.4 μg/ml of BZP after 12–15 subcultures. Lower doses of BZP (0.1 μg/ml or less) had no visible effect. Lower doses of smog were not tested. After the transitional stage of up to 15 subcultures, the transformed cultures remained transformed and normal cultures remained nontransformed for the duration of the experiments (50 subcultures). These results were obtained in triplicate sublines from each of two separate experiments and are illustrated in Fig. 1.

Periodic monitoring of the cultures by CF testing against appropriate mouse sarcoma virus-induced rat antisera showed that the virus-free cells were not reactive ($<1:2$) at the end as well as at the beginning of the experiments. The virus-carrier cultures were positive by the CF test ($>1:32$) before, as well as after, transformation occurred.

Effects on hamster embryo cultures

The HaLV-free hamster cultures were transformed after 3–6 passages under experimental conditions by 0.007 μg units but not by 0.0007 μg units of smog. The HaLV-carrier cultures were transformed after 2–4 passages by both dosages; however, the higher dose (0.007 μg units/ml) produced giant-cell formation and the lower dose (0.0007 μg units/ml) produced a jackstraw (1) type of transformation. These results were obtained in triplicate sublines from a single experiment and are illustrated in Fig. 2.

Periodic monitoring of the cultures by CF testing against appropriate HaLV-induced guinea pig antisera showed that the HaLV-carrier cultures were positive ($>1:4$) throughout the course of the experiment and that the HaLV-free culture was CF negative ($\leq1:2$) before and after transformation

77

occurred.

Tests for transplantability

These tests were made around five subcultures after transformation occurred and repeated after an additional 20–30 subcultures. None of the normal or transformed cell lines from these experiments have produced tumors in animals; however, the studies are still in progress. Many of the fully transformed cultures have been on test less than 30 days.

DISCUSSION

Under the conditions presented in these experiments, RLV-free rat embryo cultures were not transformed by smog residues or by BZP. Rauscher leukemia virus alone also did not cause transformation. Virus-carrier cultures, however, were transformed by both smog residue and BZP. These findings are entirely consistent with previous reports that rat embryo cultures were not transformed by MuLV alone or by diethylnitrosamine (1), 3-methylcholanthrene (2), or dimethylbenzanthracene (3) alone, but were transformed by combinations of MuLV and each of these three chemicals.

On the other hand, HaLV-free hamster cultures were transformed by extracts of smog residues. This is in agreement with the well-demonstrated observations of hamster embryo cell susceptibility to transformation by various chemical carcinogens (7–13). However, in our studies, the infected hamster cells were at least ten times more sensitive to transformation, which suggests that the virus genome enhanced the transformation process.

The question remains open as to whether an endogenous virus genome serves as a determinant in the chemical transformation of cells in which infectious virus cannot be demonstrated (14). In addition to hamster cultures, virus-free mouse (15) and rat embryo cultures (16) have been reported to be transformed by chemical carcinogens. However, it is often difficult to demonstrate the presence of infectious C-type RNA viruses because they do not cause cytopathic effects. Immunological reagents are often not available. The demonstration of a latent virus is even more difficult.

The concentration of smog that contained 0.0007 μg units/ml of BZP transformed rat embryo cultures as efficiently as 0.4 μg/ml of pure BZP. However, there was only one part of BZP for each 43,000 parts of total smog residue. The transforming dose of 0.0007 μg units BZP was contained in 28.5 μg of smog, which is the equivalent of 1.8 m^3 of air. We therefore conclude that the bulk of the smog residue was inactive, but contained powerful transformation-promoting agents which were approximately 600 times more efficient

78

than pure BZP.

The *in vitro* results are consistent with the findings of Kotin *et al.* who painted the skins of C57 Bl mice with atmospheric extracts (17). They found that 42% of the mice developed skin papilloma or carcinoma, beginning 465 days after treatment. Since the concentration of BZP in their preparation could not account for the number of tumors observed, they also concluded that substances other than BZP were active in promoting or causing the tumors.

The *in vitro* transformation system we describe may have definite advantages as a tool to study the effects of chemical carcinogens. Whereas Kotin *et al.* needed over a year for a single test, the *in vitro* assay system described is comparatively rapid, relatively inexpensive, easier to control, and extremely sensitive. However, the tissue culture results must be shown to correlate with *in vivo* results utilizing the same oncogenic chemicals and suitable nononcogenic analogues. It should be stressed that the usefulness of the *in vitro* assay depends not upon the ability of the transformed cells to produce tumors in newborn animals, but rather upon correlation of the visible transformation with tumor induction in animals by the same chemicals in parallel experiments. If such studies indicate that *in vitro* effects do indeed correlate with *in vivo* assays, then the *in vitro* assays should prove a useful quantitative screening test for suspected oncogenic chemicals.

There are several difficulties which prevent the utilization of the *in vitro* transformation system or the *in vivo* test system of Kotin *et al.* for estimating the danger of smog to human beings. Rat and hamster cells are very likely more sensitive to chemical oncogenesis than human cells and "virus-primed" cells are more sensitive than virus-free cells. In addition, we do not know what components of the airborne chemicals are retained in the lungs or other tissues; also, some chemicals must be altered in the body before the carcinogenic action can proceed. However, our studies provide quantitative measurements of the transforming agents in smog and also indicate that a relatively small volume of smoggy urban air may contain a tissue-culture transforming dose of smog residue.

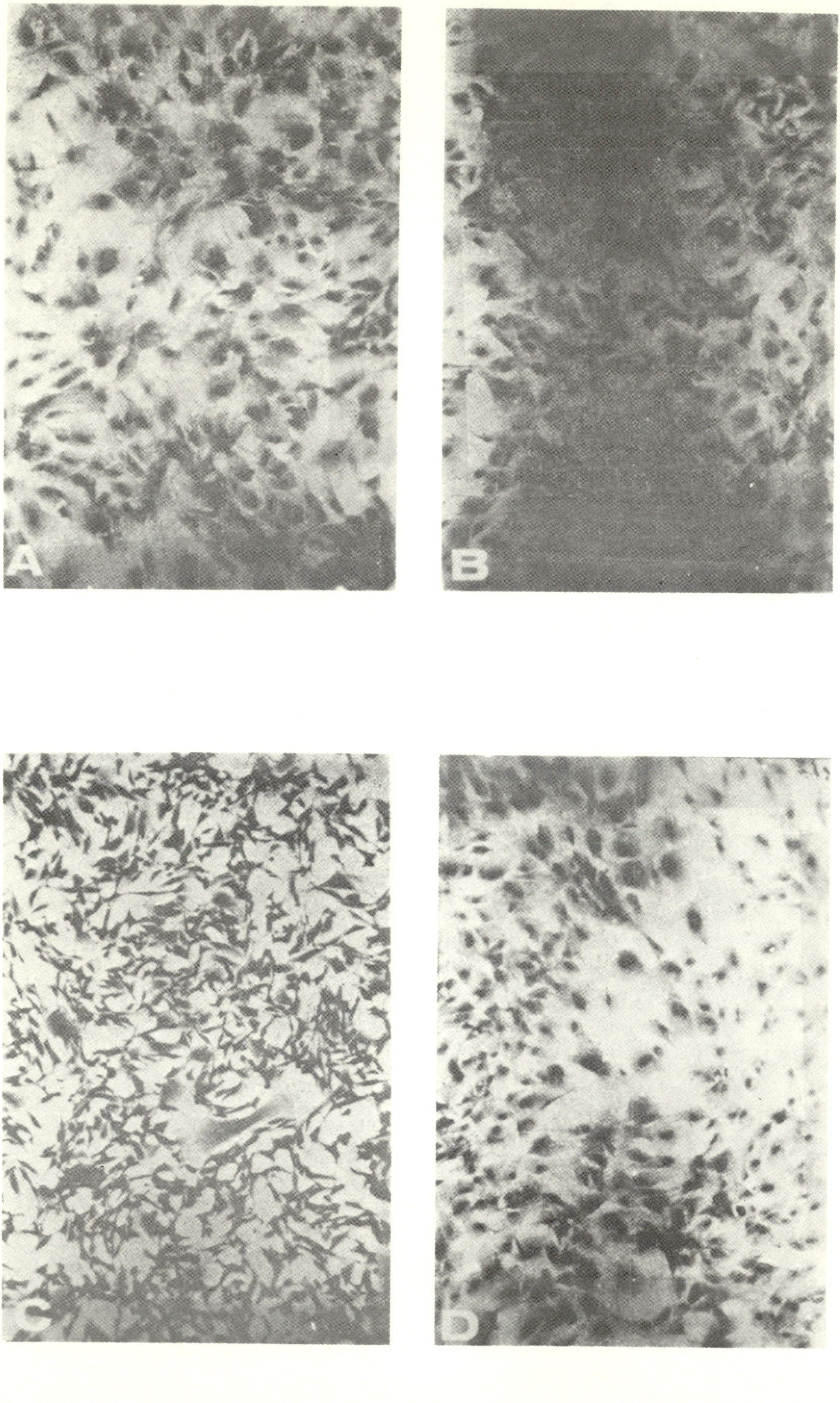
A
B
C
D

Fig. 2. Transformation of hamster embryo cells by the action of Rauscher leukemia virus
   and/or extracts of city smog.
A. LSH Hamster control (HaLV-free) P_{13} (The cultures are in a terminal stage.)
B. LSH Hamster + 0.0007 μg units/ml of city smog P_{14} (Terminal stage)
C. LSH Hamster + 0.007 μg units/ml of city smog P_{22}
D. LSH Hamster control (HaLV carrier) P_{12} (Terminal stage)
E. HaLV carrier + 0.0007 μg units/ml of city smog P_{23}
F. HaLV carrier + 0.007 μg units/ml of city smog P_{20}

All cultures were planted at 10^5 cells/ml and stained after 3 days. $\times$52. Passage number
refers to subdivision under the conditions of the experiment.   Abbreviation: LSH,
London School of Hygiene strain of Syrian hamster (Lakeview).

This work was supported in part by U.S. Public Health Contract NIH-70-2068.

1. Freeman, A. E., P. J. Price, H. J. Igel, J. C. Young, J. M. Maryak, and R. J. Huebner, *J. Nat. Cancer Inst.*, **44,** 65 (1970).

2. Price, P. J., A. E. Freeman, and R. J. Huebner, *Nature* (London), in press.

3. Rhim, J. S., W. Vass, H. Y. Cho, and R. J. Huebner, *Int. J. Cancer*, in press.

4. Hartley, J. W., W. P. Rowe, W. I. Capps, and R. J. Huebner, *J. Virol.*, **3,** 126 (1969).

5. Kelloff, G., R. J. Huebner, Y. K. Lee, R. Toni, and R. Gilden, *Proc. Nat. Acad. Sci. USA*, **65,** 310 (1970).

6. Rauscher, F. J., *J. Nat. Cancer Inst.*, **29,** 515 (1962).

7. Berwald, Y., and L. Sachs, *J. Nat. Cancer Inst.*, **35,** 641 (1965).

8. Borenfreund, E., M. Krim, F. K. Sanders, S. Sternberg, and A. Bendich, *Proc. Nat. Acad. Sci. USA*, **56,** 672 (1966).

9. DiPaolo, J. A., and P. J. Donovan, *Exp. Cell Res.*, **48,** 361 (1967).

10. Huberman, E., S. Salzberg, and L. Sachs, *Proc. Nat. Acad. Sci. USA*, **59,** 77 (1968).

11. Kuroki, T., and H. Sato, *J. Nat. Cancer Inst.*, **41,** 53 (1968).

12. Sanders, F. K., and B. O. Burford, *Nature* (London), **213,** 1171 (1967).

13. Sanders, F. K., and B. O. Burford, *Nature* (London), **220,** 448 (1968).

14. Huebner, R. J., P. S. Sarma, G. J. Kelloff, R. V. Gilden, H. Meier, D. D. Myers, and R. L. Peters, *Ann. N.Y. Acad. Sci.*, in press.

15. Chen, T. T., and C. Heidelberger, *J. Nat. Cancer Inst.*, **42,** 915 (1969).

16. Namba, M., H. Masuji, and J. Sato, *Japan. J. Exp. Med.*, **39,** 253 (1969).

17. Kotin, P., H. L. Falk, P. Mader, and M. Thomas, *AMA Arch. Ind. Health*, **9,** 153 (1954).

The effects of urban air pollution on health

An impressive body of scientific information points to the inescapable conclusion that the levels of pollutant contamination existing today in many American cities are sufficient to produce profound health consequences. This review describes the relationship between pollutant emission, atmospheric cleansing processes, and ambient air pollutant concentrations. Toxicologic studies involving the administration of sulfur dioxide, nitrogen dioxide, carbon monoxide and particulate suspensions to both animals and man are reviewed and demonstrate that single pollutants cannot explain the irritant potential of the urban atmosphere. A number of important epidemiologic studies are presented which emphasize the relationship between human illness and atmospheric pollution. Synthesis of both toxicologic and epidemologic studies leads to the conclusion that the noxious nature of the environment is due to a complicated "mix" of pollutant and meteorologic factors.

Stephen M. Ayres, M.D., and Meta E. Buehler, R.N., B.S.

The breathing of polluted air has been suspected of injuring health since coal was introduced into the English economy in the early fourteenth century. Characteristic of the public attitude to all types of health protective programs, even the most tentative action was not undertaken until a series of large-scale disasters demonstrated the urgent need for air pollution control programs. Beginning steps to regulate the use of coal were initiated in England during the reign of Richard III (1377-1399) and later under Henry V (1413-1422). That such action was not effective is seen in the essay on air pollution written by a distinguished member of the Royal Society, John Evelyn, some 250 years later. The failure to take Evelyn or his many successors seriously led to a series of air pollution episodes, producing as many as 1,063 deaths in 1909 and 4,000 deaths in the London smog of 1952.

England's role as a leader in the Industrial Revolution was inevitably followed by her role as a leader in producing air pollution. The Greater New York Area has closely followed the pattern established across the Atlantic and today ranks as the most highly polluted region in the United States. Almost nothing was done either to study or control pollutant emissions until the episodes of 1953, 1962 and 1966 produced a public reaction culminating in government action.

The effect of climate on air pollution and health

Complaints of climate influencing health are commonly expressed in the consulting room and many physicians concur with the general opinion that the northeastern part of the United States may be considered the "sinus belt." Climatologic influences may affect health in two general ways: weather factors influence the degree of air pollution, and they may also have a direct effect on human health.

The concentration of pollutant material in the ambient air depends both on the absolute quantity of pollutants emitted into the air and the quantity removed by atmospheric cleansing.[76] Wind currents are excellent cleansing agents and produce lateral movement with dilution of emitted pollutants. Vertical movement is produced by convection currents created by the temperature decline with increasing altitude. Normally, the barometric pressure decreases with increasing height above ground, allowing gas molecules to spread and leading to a decrease in gas temperature (adiabatic lapse). Since warm air tends to rise, convection currents are formed, lifting upwards earthborne pollutants.

Failure of the normal cleansing mechanisms produces an atmosphere which has a high pollution potential. High emission rates of pollutants into such an environment may produce high concentrations of pollutant material in the ambient air. In the Northeastern United States, a high pollution potential is frequently produced in the late fall and early winter by a stagnating high-pressure system or anticyclone which is associated with fair weather, low wind speeds, and temperature inversions. These periods of atmospheric stagnation lead to high concentrations of pollutants and result in "air pollution episodes" with associated injury to plants, animals, buildings, and human beings.

Weather conditions also influence the rate of pollutant production. Sulfur dioxide levels are highest in cold weather when large amounts of fuel oil and coal are consumed; they fall to low levels in the summer months.

Air contaminants found in urban air

The pollutant composition of the ambient air is closely related to economic development. The term "smog," a popular conjugation of "smoke" and "fog," was used to describe the English particulate-laden environment produced by the universal burning of coal. This type of air pollution, produced by various-sized particulates and sulfur oxides, is characterized by its chemical reducing action and is the dominant type of atmospheric pollution found in large industrialized areas such as New York, Philadelphia, and Chicago where coal and other fossil fuels are consumed in large quantities.

Another major type of air pollution is

the well-known Los Angeles smog which, although irritating eyes, decreasing visibility, and injuring plants, contains much lower concentrations of sulfur oxides and particulates than the reducing smog described above. Professor A. J. Haagen-Smit[51] demonstrated that ultraviolet irradiation of mixtures of hydrocarbons and nitrogen oxides produced ozone and other oxidants, leading to a smog similar in biologic properties to that occurring naturally in Los Angeles. Subsequent studies have shown characteristic diurnal variations. Hydrocarbon and nitrogen oxide levels are high early in the day corresponding with the peak use of automobiles, while oxidant levels are at their maximum several hours later reflecting the delayed effect of sunlight on the hydrocarbon–nitrogen oxide mixture. Photochemical smog accumulates in congested areas with high densities of automobiles, low wind speeds, temperature inversions, and adequate sunlight to permit photochemical transformation. While initially recognized in the Los Angeles area, photochemical smog has now been observed in every region of the United States.

A detailed account of photochemical atmospheric reactions has recently been published.[52] Briefly, the most important primary photochemical reaction appears to be the photodissociation of nitrogen dioxide into nitric oxide:

$$NO_2 + hv = NO + O$$

where hv is the symbol for a light energy unit. Another important reaction is the photodissociation of aldehydes into free radicals:

$$RCHO + hv = \dot{R} + H\dot{C}O$$

Ozone may be formed by the reaction of free and molecular oxygen:

$$O + O_2 = O_3$$

or various peroxy compounds formed by combination of the free radical with oxygen:

$$\dot{R} + O_2 = R\dot{O}O$$

The peroxy radicals react further with nitrogen oxides and other pollutant substances to produce a number of secondary substances including alkyl nitrates, peroxyacyl nitrates, alcohols, ethers, acids, and peroxyacids. An important peroxyacyl nitrate, peroxyacetyl nitrate (PAN) exerts a specific toxic effect on plants which may be used as an indicator for its presence.

Table I lists average urban air based on 1963 figures.[84] Although a large number of organic and inorganic compounds are found in urban air, most studies of pollution-related health effects have monitored but several of these. A health correlation based on the observed concentrations of sulfur dioxide might be strengthened or weakened if based on the concentration of another pollutant. Obviously the ideal approach would be to measure a large number of pollutants and biologic effects and to interrelate them by statistical techniques.

Measurement of air pollutants and relative concentrations in various areas

An understanding of measurement techniques and levels is important for interpretation of the health effects to be described in the following sections.

Sulfur and sulfur compounds. Oxidized sulfur, a product of combustion of fossil fuel, has been extensively studied and related causally to health effects. Sulfur dioxide is the major form found in the atmosphere although smaller amounts of nonvolatile sulfuric acid and sulfate salts are also present. Sulfur dioxide, measured in parts per million, is probably the most widely measured pollutant. Unfortunately,

Table I. *"Average" urban air, 1963, approximate composition*

Compound[a]	Concentration ($\mu g/1,000$ M.^{3})
Carbon dioxide	6.3×10^8
Carbon monoxide	8×10^6
Methane	1×10^6
Ethylene	1×10^5
Benzene	1×10^5
Airborne particulates	1×10^5
Sulfur dioxide	8×10^4
Formaldehyde solution	7×10^4
Nitrogen dioxide	6×10^4
Nitric oxide	4×10^4
Phenols	2×10^4
Ammonia	2×10^4
Oxidants (as ozone)	2×10^4
Particulate acid (as H_2SO_4)	1.4×10^4
Acrolein	1.4×10^4
Sulfates	1×10^4
Benzene-soluble particulates	7×10^3
Large aliphatic hydrocarbons	3×10^3
Nitrates	2×10^3
Iron	2×10^3
Lead	8×10^2
Large n-alkanes (C_{14} to C_{30})	5×10^2
Zinc	2.3×10^2
n-Tricosane	8×10^1
Manganese	7×10^1
Copper	6×10^1
Titanium	3×10^1
Nickel	3×10^1
Arsenic	2×10^1
Tin	2×10^1
Vanadium	2×10^1
Chromium	1×10^1
7H-Benz(de)anthracen-7-one	8
3.4-Benzpyrene	6
Phenalen-1-one	2
Benz(c)acridine	1
Dibenz(a.h)acridine	2×10^{-1}
Phenols/3,4-benzpyrene	3,300
Aliphatic hydrocarbons/ 3.4-benzpyrene	500
n-Alkanes/3,4-benzpyrene	83

the various methods vary in specificity, and the concentration measured by one technique may not be identical with that measured by another.[75]

Particulate sulfates are measured by high-volume sampling devices which force air through filter paper. The filtered particles may be weighed, sized, and analyzed by chemical methods. Suspended sulfate measured by this technique averages 5 to 15 μg per cubic meter. Total sulfation rates may also be determined by the lead peroxide candle method in which lead paste is converted to lead sulfate over varying periods of time. Total sulfation is usually reported in units of milligrams of sulfate per 100 cm.^{2} of exposed lead peroxide candle per day.

Particulate pollution. A number of techniques for analysis of particulate pollution are in current use. Unfortunately it is frequently impossible to extrapolate data obtained with one technique to another.[74] Dustfall is the measure of particles which settle and generally is an index of particles greater than 10 microns in size. These particles are not respirable and the dustfall measurements do not correlate well with biologic effect. Three other techniques are in widespread use and are generally selected for health studies.

Particles may be sized and counted in a photoelectric device which consists of a light source designed to illuminate a small volume of sample and a photomultiplier tube to detect flashes of light reflected by the individual particles. Perhaps the most widely used devices are tape samplers which draw air through successive areas of filter paper for a given period of time. The amount of particulate material filtered is measured by either reflectance or transmittance techniques. The former is widely used in the United

States and the measurements are expressed as coefficient of haze (COH) units per thousand linear feet of air. The high volume sampler is used to measure total particulate concentration in large volumes of air. A major advantage is that the particles may be separated and analyzed chemically for various components.

Photochemical smog. The intensity of photochemical smog may be indicated by the measurement of total oxidants, ozone, or certain organic oxidants such as peroxyacetyl nitrate (PAN).

Other pollutants. Carbon monoxide is commonly measured by an infrared gas analyzer and reported in parts per million. Specialized techniques exist for the measurement of oxides of nitrogen, hydrocarbons, and other pollutants.[91]

Average values and seasonal variations in urban areas. Figs. 1 and 2 represent soiling index (COH), particulate concentration, and sulfur dioxide and nitrogen dioxide concentrations found in New York City during 1967. Note the marked decrease in sulfur compounds and particulates in the summer months reflecting a decrease in fuel consumption, while nitrogen oxides produced primarily by automobile exhaust remain relatively constant. Note that while the two measures of particulate pollution correlate reasonably well, occasional discrepancies occur.

The time of sampling and averaging is extremely important. Both monthly maximum 24 hour averages and monthly averages are presented in Figs. 1 and 2. A real problem exists in the presentation of air quality data. A low annual average may conceal a peak period of extremely high air pollution and health effects may well be sensitive to extremely high peaks as well as to average exposure. Most monitoring networks report hourly, daily, and

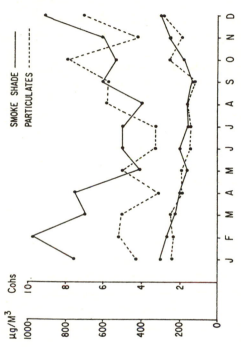

Fig. 1. Smoke shade (COH units) and suspended particulates (micrograms per cubic meter) for New York City in 1967. The upper two lines represent 24 hour maximums while the lower two lines represent monthly averages. Note that the 24 hour maximums may be considerably higher than monthly averages. There is a striking seasonal variation with peak concentrations observed in the winter months.

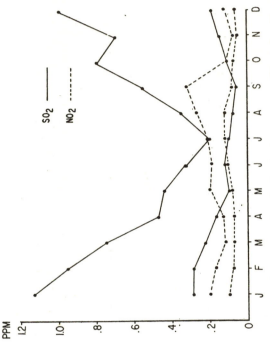

Fig. 2. Sulfur dioxide and nitrogen dioxide levels (parts per million) for New York City in 1967. Upper two lines represent 24 hour maximums and lower two lines monthly averages as in Fig. 1. Sulfur dioxide concentrations show marked seasonal variation and considerably higher daily maximums compared to monthly means. In contrast, nitrogen dioxide, produced primarily by automotive exhaust, is relatively stable throughout the year and does not show the striking difference between 24 hour maximums and monthly means.

yearly averages. Individual 24 hour maximum averages may vary from annual means by as much as 300 to 700 per cent.

Table II lists yearly mean values and maximum 24 hour values for certain gaseous pollutants and suspended particulates for a number of American cities.[40]

The effects of air pollution on health

Characterization of the specific effects of urban air pollution on human health is based on evaluation of a series of epidemiologic, toxicologic, and physiologic studies. Casual observations suggesting that London coal smoke might lead to respiratory impairment have formed the basis for epidemiologic studies clearly relating certain types of urban air pollution to respiratory disease. Limited studies of controlled human exposure, buttressed by a number of detailed animal exposure studies, have attempted to evaluate the irritant and toxic potential of specific environmental pollutants. Air pollution control—the ultimate aim of health effects studies—is based on the establishment of air quality standards derived from a synthesis of epidemiologic and toxicologic data.

Toxicologic investigations into the effects of air pollutants

Epidemiologic investigations to be reviewed in the next section have stimulated toxicologic studies of the physiologic effects of individual pollutants. While important information has been obtained from these investigations, it has generally been impossible to reproduce human health effects in a laboratory environment in concentrations similar to those existing in polluted urban air. This discrepancy between toxicologic and epidemiologic study sug-

Table II. Pollutant levels in selected cities

	Chicago	Cincinnati	Philadelphia	Denver	St. Louis	Washington, D. C.
Sulfur dioxide	0.08 (1.14)	0.03 (1.86)	0.09 (0.87)	0.01 (0.96)	0.04 (1.25)	0.04 (0.47)
Nitric oxide	0.10 (0.74)	0.04 (1.18)	0.06 (1.98)	0.04 (0.59)	0.03 (0.61)	0.04 (1,15)
Nitrogen dioxide	0.06 (0.35)	0.04 (1.59)	0.04 (0.29)	0.03 (0.35)	0.03 (0.21)	0.04 (0.19)
Carbon monoxide	13.0 (66.0)	5.0 (32.0)	7.0 (47.0)	8.0 (63.0)	6.0 (68.0)	3.0 (47.0)
Total oxidants	0.02 (0.23)	0.02 (0.13)	0.03 (0.60)	0.03 (0.30)	0.04 (0.25)	0.03 (0.22)
Hydrocarbons	2.8 (14.9)	2.7 (12.8)	2.5 (14.4)	2.4 (19.1)	3.0 (14.3)	2.4 (14.5)
Suspended particulates	140.0 (244.0)	141.0 (255.0)	182.0 (312.0)	156.0 (510.0)	152.0 (260.0)	98.0 (199.0)

First value listed is annual mean. Values in parentheses are 24 hour maximums. All are reported in parts per million (p.p.m.) except for suspended particulates which are reported in micrograms per cubic meter (μg/M.^{3}).

gests that the total irritant potential of polluted air is related to a mixture of air contaminants and to associated unfavorable weather conditions such as low temperatures, high barometric pressures, and high humidity.

The toxicologic literature pertaining to air pollution is large and this review will merely sample several particularly pertinent areas of study.

Effects of deliberate exposures to sulfur dioxide. An example of toxicologic design and interpretive difficulties is the study of the effects of inhaled sulfur dioxide on respiratory function. One of the earliest toxicologic studies was that of Amdur and associates[7] who administered sulfur dioxide to 14 normal subjects and observed an increase in pulse and respiratory rate and a decrease in tidal volume when exposed to sulfur dioxide at a concentration of 5 p.p.m. for ten minutes. Changes were also seen at levels as low as 1 p.p.m. Lawther[65] was unable to confirm these findings and suggested that they were due to chance observations in a small series of individuals.

A detailed series of experiments evaluating the effect of various agents on pulmonary flow resistance in the guinea pig was then undertaken by Amdur with the use of a technique developed by her and Mead.[6] A summary of her sulfur dioxide studies discussed against the background of sulfur dioxide absorption studies of Strandberg has been recently published.[4] Amdur plotted dose-response curves with the use of a one hour exposure to various concentrations of sulfur dioxide. A biphasic effect was observed with an increase in airway resistance occurring with as little as 0.16 p.p.m. but greater proportional effects being observed at concentrations above 20 p.p.m. Strandberg[92] had shown that the efficiency of upper respiratory scrubbing and pollutant removal was much greater with high concentrations of SO_2, so that greater percentages of lower concentrations reached the lower airway. Amdur took Strandberg's absorption data, applied them as a correction factor to her own data, and observed a reasonably straight dose-response curve, over the range 0.16 to 835 p.p.m. of SO_2.

Human exposure studies have been published by a number of investigators with varying results. Frank and associates[36] demonstrated increases in airway resistance in 11 healthy subjects when 5 and 13 p.p.m. of SO_2 were breathed for 5 to 30 minutes. While the two higher concentrations increased airway resistance, only one of the subjects experienced an increase in airway resistance with the lowest concentration. Snell and Luchsinger[87] were able to construct a dose-response curve by administering 0.5, 1.0, and 5.0 p.p.m. for 15 minutes with the use of maximum expiratory flow rates at half lung volume as an indicator. Five of the 9 subjects had decreased flow rates at the lowest concentration but the changes were not significant for the entire group. Significant changes were observed following breathing of either 1 or 5 p.p.m. of SO_2. Speizer and Frank[88] found that 15 or 28 p.p.m. of SO_2 breathed for 10 minutes significantly increased airway resistance but that the effects were substantially greater when the gas was mouth breathed rather than administered by nasal mask.

In contrast to these studies, Wright[106] did not observe consistent changes in airway resistance following breathing of 2.5 to 23 p.p.m. of SO_2 for 20 minutes, and Burton and associates[20] failed to alter airway resistance when their 10 subjects breathed 2 to 3 p.p.m. for 30 minutes.

o

Partial explanation for the apparent inconsistency in these studies is found in our own published data. We exposed 96 subjects with or without pulmonary disease to a high concentration (50 p.p.m.) of SO_2 for a short period (10 breaths) in an attempt to determine the variability of response. Considerable variation in response was found although the patients with bronchitis and asthma had significantly greater responses, as a group, than did the normal subjects. This individual variation may explain the inconsistency in reported studies. The observation that bronchitic and asthmatic patients have greater responsiveness to inhaled pollutants confirms Amdur's findings that guinea pigs with high resistances appeared more sensitive to airway irritants[3] and also suggests that certain individuals in an exposed population may be more likely to suffer health effects during periods of high pollution.

Although statistical changes in pulmonary function are found following inhalation of sulfur dioxide, it is difficult to imagine how community air pollution with average levels of 0.1 to 0.4 p.p.m. of SO_2 and occasional peaks to 1.0 to 1.5 p.p.m. can produce the significant health effects suggested by epidemiologic studies. The inability to reproduce clinical effects at levels encountered in the urban atmosphere has focused attention on other components—particularly particulate material—found in polluted environments.

Nitrogen oxides. Nitrogen oxides are important toxic constituents of both automotive exhaust and tobacco smoke, are precursors of ozone under conditions promoting photochemical transformation, and are believed to exert significant independent effects on pulmonary tissue. The acute pulmonary insufficiency seen in farmers working in silos, silo-fillers' disease, is believed related to inhalation of high concentrations of nitrogen dioxide and provides a human experimental model of the consequences of exposure to that substance. This pollutant may well explain part of the relationship between chronic obstructive pulmonary disease and smoking, since tobacco smoke contains approximately 250 parts per million of nitrogen dioxide.[93] Table II and Fig. 2 show that the concentration of nitrogen dioxide in urban air is usually less than 1 p.p.m.

Freeman and associates[37] have published a series of studies reporting the effects of various concentrations of nitrogen dioxide on the rat lung. Rats exposed continuously to 0.8 p.p.m. NO_2 survived natural lifetimes but consistently exhibited mild tachypnea. Continuous exposure to 2 p.p.m. NO_2 produced changes in the terminal bronchiole and alveolar duct with broadening of bronchiolar epithelial cells and loss of cilia. Inclusion bodies within lining cells suggested deficient cleansing of inhaled or metabolic waste material. Exposure to 10 to 25 p.p.m. produced large, air-filled heavy lungs without pulmonary edema which resembled those in human emphysema and which produced death from respiratory failure in 20 to 22 weeks. Microscopic study revealed narrowing of the terminal bronchiolar lumens due in part to epithelial cell hypertrophy and in part to accumulation of cellular and noncellular debris at the junction of alveolar duct and bronchiole. Distention and fragmentation of alveoli resembled that seen in human emphysema. Of interest, alveolar lining cells were hypertrophied and appeared to compress septal capillaries presenting an additional barrier to gas exchange and presumably leading to the uniformly observed polycythemia.

Respiratory mechanics and alveolar gas exchange were studied in rabbits continuously exposed to 8 to 12 p.p.m. of NO_2 for 3 months by Davidson and associates.[28] Functional residual capacity and airway resistance increased, static lung compliance was unchanged, and arterial oxygen tension decreased. Arterial carbon dioxide tension was unchanged.

Boren[17] studied the interaction of nitrogen dioxide with carbon particles in an attempt to approximate part of the "pollutant mix" found in the community atmosphere. Exposure of mice to 250 to 500 p.p.m. NO_2 produced pulmonary edema and frequent death while exposure to carbon particles alone or daily exposures to 25 p.p.m. of NO_2 for 30 minutes was without effect. Exposure to carbon particles previously treated with NO_2 produced focal destructive lesions with loss of alveolar walls.

Behavior of inhaled particles. While inhaled gases may move rapidly in and out of the respiratory system, the removal of inhaled particles is considerably slower and depends upon the efficiency of the tracheobronchial mucociliary carpet and of alveolar macrophage activity. The physiologic response produced by an inhaled gas is related to its solubility.[25] Sulfur dioxide, a highly soluble gas, may produce an immediate bronchoconstrictor effect because it is immediately absorbed by upper respiratory receptors; nitrogen dioxide is considerably less soluble, may reach more distant lung regions, and produce delayed nonreflex effects. The physiologic response of a particle, in contrast, is determined by its chemical composition and size.

The most meaningful characterization of particle dimension is the equivalent aerodynamic diameter which is the diameter of a sphere of density 1 gm. per cubic centimeter which falls in air with the same terminal settling velocity as the particle in question. Hatch and Hemeon[55] showed that the alveolar retention of particles is related to two basic processes, particle deposition and particle clearance. Particles greater than 3 to 5 μ in size tend to be deposited in the upper respiratory tract while alveolar deposition increases with decreasing particle sizes. Particle clearance, on the other hand, is more efficient for smaller particles. The net result of these conflicting factors is to produce a zone of maximal alveolar retention which varies among particle species but ranges from 0.5 to 3 μ. The effect of particle species on retention is seen in the study of Landahl and Herrmann[64] who found that the particle diameter corresponding to 50 per cent retention ranged from 0.9 μ for $NaHCO_3$ to 3 μ for corn oil.

Wilson and LaMer[102] confirmed theoretic predictions in human subjects by measuring alveolar and total retention following inhalation of a radioactive aerosol ($Na^{24}Cl$). Peaks of maximal retention at diameters ranging from 0.3 to 1.0 μ were observed. Prolonged particle persistence was demonstrated by Albert and co-workers[1] who administered radioisotope-tagged iron oxide particles of various sizes to human subjects and followed chest radioactivity with sodium iodide scintillation counters. In one experiment, 30 per cent of 3.6 μ particles remained in the lung after 56 days. In another study, about 30 per cent of 5.0 μ particles and 65 per cent of 2.9 μ particles were retained at 24 hours.

The combined effects of particle size and composition are shown in the study of Amdur and Corn.[5] The percentage increases in guinea pig airway resistance produced by inhalation of similar concentrations of zinc ammonium sulfate[1] were 80 per cent, zinc sulfate 40 per cent, and ammonium sulfate 25 per cent. Dose-response curves were then prepared for increasing concentrations of particulate sulfate at different particle sizes. The slopes of the different curves were dramatically different; increasing concentration of particles of 1.4 μ diameter produced small increments in airway resistance while increasing concentrations of particles of 0.3 μ produced dramatic increments in resistance. Equivalent concentrations of these two particle sizes produced 5 per cent and 80 per cent increases in airway resistance, respectively.

Zinc ammonium sulfate had been selected by Amdur and Corn because earlier studies suggested its possible irritant role in the acute air pollution episode occurring in Donora, Pennsylvania. The relevancy of the particle sizes investigated was later demonstrated by Corn and DeMaio[26] who found that 90 per cent by number of the total particulate sulfates in Pittsburgh air were less than 1.9 μ in diameter.

The studies reviewed in this section have given rise to the concept of "respirable" air pollutants. While measures of total suspended particulates are widely available, health effects may be more closely related to the concentration of respirable particulates. The chemical composition of these two groups of particulates may vary widely. Benzene soluble organic substances, total sulfates, and iron have been shown to make up 7.7, 15.3, and 5.8 per cent of respirable dust and 3.5, 9.2, and 17.3 per cent of total suspended particulate material.

Gas-particle interactions. The ubiquitous occurence of certain pollutant gases and the demonstrated retention and irritability of small particles have led to the concept of gas-particle interactions. A particle might potentiate the effect of an absorbed gas by promoting deposition or by providing a surface for catalytic transformation of the gas to a more toxic form. Amdur[2] showed that an 0.04 μ aerosol of sodium chloride potentiated the effect of sulfur dioxide on the guinea pig while a 2.5 μ aerosol did not. Zinc sulfate particles 0.29 μ in diameter were shown by Amdur and Corn to potentiate the effect of sulfur dioxide. Carbon particles were shown by Boren[17] to act as a carrier for nitrogen dioxide producing focal destructive pulmonary lesions reminiscent of human emphysema, while administration of nitrogen dioxide alone produced pulmonary edema.

Several investigators have attempted to reproduce the animal observations of Amdur and associates in man. While neither Frank and associates[35] nor Burton and associates[20] could demonstrate an increased effect with a sulfur dioxide aerosol mixture, Snell and Luchsinger[87] found that a distilled water aerosol (mean particle size about 0.3 μ) augmented the response to sulfur dioxide while a saline aerosol (mean particle size about 6 to 8 μ) did not.

Experimental observations of breathing filtered air. While it may be ethically impossible to perform certain toxicologic studies, the reverse procedure—elimination of pollutants from an urban atmosphere—is both ethically proper and scientifically rewarding. Motley and associates[73] demonstrated significant improvement in pulmonary function studies in emphysematous patients who breathed filtered air for at least 40 hours. Recently Ury and Hexter[97] analyzed an experiment performed by Remmers and Balchum and were able to demonstrate a correlation between pulmonary function tests and pollutant levels in a group of 15 patients with severe obstructive emphysema who lived for 3 weeks in a controlled environment room.

Biologic effects of photochemical smog. Toxicologic studies of the products of photochemical smog on human subjects stand in a different position than those on the gases and particulates discussed above because definite laboratory effects may be observed at levels occuring in community atmospheres. Most toxicologic studies have measured either total oxidant or ozone concentrations. Since ozone may constitute up to 90 per cent or more of the total oxidant level, the two measures may be roughly equated. The studies described below should be analyzed with regard to the fact that total oxidant levels in certain communities may average 0.2 to 0.5 daily and may peak to levels of 1.0 p.p.m. in dense photochemical smog.

Eye irritation is experienced when total oxidant levels reach 0.1 to 0.2 p.p.m. and concentrations of ozone between 0.2 to 0.5 p.p.m. reduce visual acuity. Brief exposure to 0.05 to 0.1 p.p.m. produces irritation and dryness of the upper respiratory passages and concentrations between 0.30 to 1.0 produce choking, coughing, and severe fatigue.[61] Diffusing capacity and other pulmonary function studies were significantly decreased in normal subjects exposed to 0.6 to 0.8 p.p.m. ozone for 2 hour periods,[108] and the decrease in function persisted for up to 24 hours. Griswold and associates[49] exposed a normal volunteer to 1.5 to 2.0 p.p.m. and produced a syndrome characterized by impaired lung function, severe chest pains, coughing, headache, and difficulty in coordinating which persisted for almost 2 weeks.

Epidemiologic studies

Goldsmith[42] has emphasized that epidemiology is literally translated as "the science of that which is upon the people." He pointed out that the total management of environmental problems from a general concern for disease states, through identification of causative environmental factors, to the establishment and observation of workable control systems is basically the responsibility of the epidemiologist. This review will outline certain epidemiologic principles and present data from a number of important studies attempting to evaluate the relationship between urban air pollution and health.

Air pollution epidemiology attempts to relate various air phenomena with various health phenomena. Health phenomena may be observed by recording general or specific death rate or by various indices of morbidity including clinic visits, absenteeism in working populations, observation of panels of selected patients, or interview techniques on various population groups. Physiologic effects which may or may not produce symptoms may be identified by the application of physiologic testing procedures or by pathologic examination of autopsy or biopsy material. Air phenomena are evaluated by meteorologic pollutant concentration and pollutant emission studies. The total body burden of certain air pollutants may be determined by frequent air quality measurements or predicted from mathematic models with the use of emission and meteorologic data.

Interpretation of air pollution epidemiologic data is facilitated by the suggestion of Ipsen and associates[59] that air and health phenomena may be interrelated in at least four ways. The obvious occurrence of both high levels of air pollution and excessive mortality and morbidity rates has been called the "double phenomena" and is typified by the classic air pollution episodes. The obvious outbreak of a specific disease (epidemic) may be related to a recognized air contaminant, an example of specific epidemiology, or a unique air phenomenon may be studied in retrospect and related to more subtle health phenomena. A fourth interrelationship is the statistical association of subtle air and health effects over a definite time period.

Quantitative study of the above relationships requires accurate measurement of both air and health phenomena and definition of the relationship between these two factors. A number of considerations may confuse the relationships between air pollution and health effects. The unmeasured variable may either produce an apparent relationship or may obscure an actual relationship. Frequency of cigarette smoking, genetic background, climate, presence of influenza virus or other respiratory infectious agent, and economic level are other variables which may alter the frequency of health phenomena. Both air pollution and health effects might be related to a third variable and not themselves interrelated. For example, cold weather might produce upper respiratory infections and also increase fuel consumption producing an increase in air pollution. A study correlating upper respiratory infections and air pollution would predict a causal effect where, in fact, none might exist. If the effect of cold temperature could be controlled by making a series of health observations during two periods—both with cold temperature but one with high pollution, and the other with low pollution—more meaningful data might be obtained. Many of the studies reviewed below have attempted to control certain of these factors by evaluating populations in different geographic locations, and different areas of the same urban area and children (presumably non-smokers) or by measuring all conceivable variables and eliminating them statistically.

The conclusions of most epidemiologic studies are based on statistical techniques ranging from simple standard testing procedures to complicated analyses made possible by high-speed digital computers. Unfortunately for many physicians, the statistical techniques used to support a given conclusion are frequently unfamiliar and reliance must be placed on the statistical ability of the editorial review. The papers of Ury and Hexter,[97] Reinke,[82] and Sterling and associates[90] should be consulted for an introduction to air pollution statistics. Standard analysis of variance and simple correlation procedures may be used when the data are normally distributed. Nonparametric tests such as chi square, sign test, or Wilcoxon tests may be used if the data are not normally distributed. Multiple regression with either linear or nonlinear treatment is especially useful when evaluating the effect of multiple factors on various health indices. Verma and associates[98] related both health and environmental variables to time and attempted to remove time trends in order to study the relationship between health effects and air pollution. This study and the series of papers by McCarroll and associates[70] make use of a series of correlations for each time interval. The latter authors, for example, published "correlograms" which graphically presented both cross- and autocorrelations for a group of symptoms and a group of air pollutants.

Air pollution episodes: The double phenomena. Certain striking episodes characterized by obvious increase in air pollution and obvious health effects provided

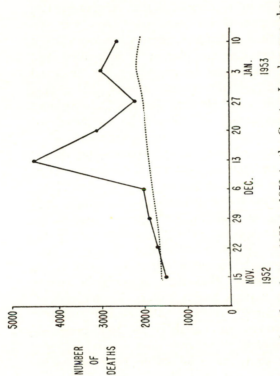

Fig. 3. Daily mortality data during 1952 to 1953 in the Greater London area showing the striking increase in death rate during a period of extremely high pollution. Dashed line shows daily mortality data for the period 1947 to 1951 and the difference between the two lines is considered "excess mortality." (Redrawn from Goldsmith.[41])

the earliest suggestion that air quality and health might be closely related. This section deals with the major air pollution disasters (double phenomena of Ipsen[59]) and also reviews certain other episodes of high pollution which were retrospectively associated with increased death and morbidity. Obviously, as public and professional awareness increases, more such episodes may be regarded as double episodes.

On Monday, December 1, 1930, a thermal inversion markedly reduced the diluting capacity of the atmosphere in the Meuse River Valley in Belgium, a river valley 15 miles in length surrounded by hills rising to 300 feet on either side. Chest pain, cough, shortness of breath, and eye and nasal irritation were common symptoms; 60 persons died in a period of one week. Air quality measurements were not made but it is estimated that sulfur dioxide concentrations ranged between 10 and 38 p.p.m. Goldsmith[41] has suggested that sulfuric acid, produced by combination of sulfur dioxide with water vapor in the presence of other pollutants, was probably one of the major irritating substances present in the atmosphere.

The effect of a similar atmospheric inversion occurring in Donora, Pennsylvania, located in a valley of the Monongahela River, has been masterfully portrayed by Berton Roueché[53] in his book, *Eleven blue men: Tales of medical detection.* The episode began on the morning of Tuesday, October 26, 1948. Over 40 per cent of the population became ill with lower and upper respiratory symptoms; 20 died, most on the third day of the episode. While the pollutants responsible for the irritant effects have not been identified—measurements were not made during the episode— sulfur dioxide concentrations were estimated to be as high as 2.0 p.p.m.[41]

The most disastrous air pollution episode occurred in the British Isles from December 5 to 9, 1952. Fig. 3 shows the number of deaths registered in Greater London for each week beginning with November 15 and continuing through January 10. The dashed line shows the average death rate for the period of 1947 through 1951; deaths above the dashed line represent "excess" deaths. The total excess was between 3,500 and 4,000 deaths for the entire period. Even though atmospheric conditions improved within 4 days, the increase in deaths persisted throughout the month of December, suggesting that the initial pollutant insult was complicated by other factors (presumably infection) which perpetuated the mortality rate.

A number of periods of high air pollution have been recorded in New York City—a city which consistently experiences extremely high levels of atmospheric contaminants. A period of persistent temperature inversions occurred from November 15 to November 24, 1953. The intensity of air pollution was seen in the increase in the daily average smoke shade readings from a peak (for the period of observation) of 2.42 in 1950 through 1952 and 2.76 in 1954 through 1956 to a peak of 8.38 on November 20, 1953. Greenburg and associates[48] compared the daily death rate during that period with that observed in 1950 through 1952 and 1954 through 1956. The daily number of deaths was greater than 250 in only 12 of the 210 control days (5.7 per cent), while that number was exceeded on 6 of the 10 days (60 per cent) of the November, 1953, inversion period. An increase in clinic visits for upper respiratory and cardiac illnesses but not for asthma was also observed during this period.[47]

Many of the New York City episodes have not been true "double phenomena," in which both health and air phenomena were immediately apparent. The subtle effects on human health may be seen in a series of statistical studies conducted by Greenburg and associates[45, 46] and by McCarroll and Bradley.[69] Fig. 4 is taken from the latter investigators and demonstrates the variety of interrelationships observed between human health and the environment. Daily deaths from all causes, sulfur dioxide and smoke shade concentrations, are shown. Two types of control mortality data are shown. The solid line shows the expected daily mortality rate based on data from previous years; the dotted line is a 15 day moving average based on the 7 days preceding and following the day in question.

November 26 to December 6, 1962. Sudden increases in daily death rate and air pollution which appear to be related are observed. An increase in morbidity among individuals living in nursing homes was also observed during this period.[45] McCarroll and Bradley[69] pointed out that while mortality rate tends to fall below expected levels following such a sudden rise in pollutant levels, this fall is never sufficient to compensate for the excess of deaths on the preceding days.

December 30, 1962 to January 15, 1963. This period demonstrates certain other environmental relationships confusing the direct effect of air pollution on health. A sudden temperature drop occurred on the night of December 30 to below 10° F. A broad peak of increased mortality rate follows this but two other mortality rate peaks appear related to air pollution peaks during a period of relatively constant temperature.

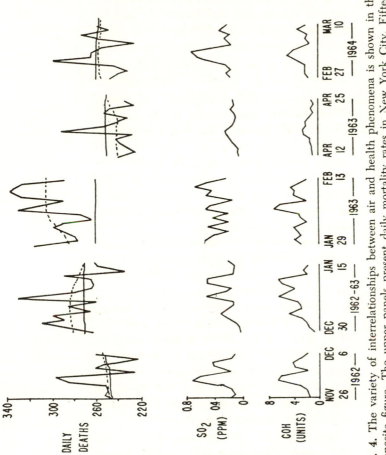

Fig. 4. The variety of interrelationships between air and health phenomena is shown in this composite figure. The upper panels present daily mortality rates in New York City. Fifteen day moving averages are shown in dotted lines and, predicted mortality rate based on previous years is shown in solid lines. Sulfur dioxide and smoke shade concentrations are presented as indicators of air pollution. See text for complete description. (Adapted from McCarroll and Bradley.[69])

January 29 to February 13, 1963. The presence of an influenza epidemic may seriously distort mortality rate figures. The daily mortality rate is well above predicted levels and the moving average is obviously a more appropriate base line. Air pollution and mortality rate peaks appear to be related even though the mortality figures are considerably higher than usual.

April 15 to April 25, 1963. This period demonstrates the presence of an increased mortality rate peak without increase in the air pollution indices but during a period of decreased wind speed and increased frequency of inversions. The mortality rate peak is statistically unexpected and difficult to explain. The sudden rise is not compatible with an infectious disease process. Other unmeasured pollutants may have been elevated or other yet undetermined environmental factors may have been responsible. Cold temperature and influenza virus could be eliminated as possible causal factors.

February 27 to March 10, 1964. This episode was characterized by extremely high levels of air pollution which suddenly appeared and were associated with extremely low wind speeds and an increased frequency of temperature inversion. It appears to be an inescapable conclusion that the sudden change in concentration of air pollutants from low levels to very high levels is responsible for the sharp increase in mortality rate.

The Thanksgiving Day air pollution episode. On November 19, 1966, a large mass of cold air—an anticyclone—moved over the Eastern seaboard and persisted until November 25. Wind velocity fell to 4.3 miles per hour and daily temperature inversions appeared. Fig. 5A shows hourly concentrations of carbon monoxide, smoke

112

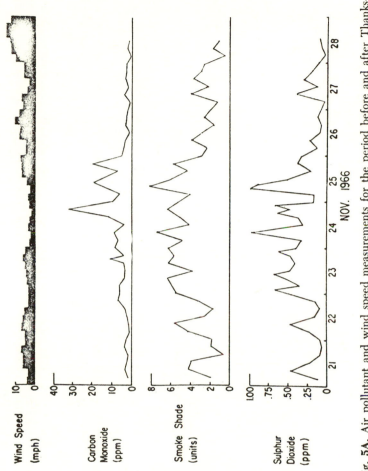

Fig. 5A. Air pollutant and wind speed measurements for the period before and after Thanksgiving Day, 1966. Thermal inversions and decreased wind speeds created a stable atmosphere with striking increases in carbon monoxide, smoke shade, and sulfur dioxide.

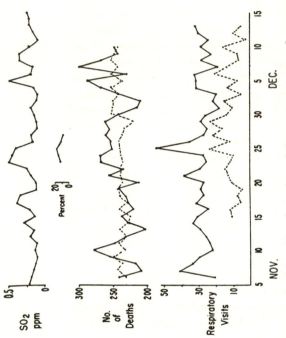

Fig. 5B. Four separate health indices from three different investigators demonstrate the effects of the Thanksgiving Day inversion on human health. From above downward, the top panel shows sulfur dioxide concentrations while the second panel shows the incidence of eye irritation during a five day period studied by Becker, Schilling, and Verma.[15] The middle panel shows daily mortality data together with predicted mortality (*dashed line*) from Glasser, Greenburg, and Field.[39] The two lower panels show the pattern of emergency room visits for respiratory complaints from Glasser, Greenburg, and Field[39] together with our own unpublished data (*dashed line*).

shade, and sulfur dioxide. The intensity of the pollution is shown by the hourly maximum for sulfur dioxide which reached 0.69, 0.97, and 1.02 on November 23, 24, and 25, respectively. Pollutant concentrations dropped precipitiously on the evening of November 25 due to a sudden change in meteorologic conditions with increasing wind speeds.

Four separate health indices are presented in Fig. 5B. Pollution levels are indicated by 24 hour average values for sulfur dioxide. The second panel shows the number of positive responses to a query regarding eye irritation contained in a questionnaire administered by Becker, Schilling, and Verma.[15] The middle panel shows daily mortality data together with predicted mortality rate (dashed line) from Glasser, Greenburg, and Field.[39] The two lower panels show the frequency of emergency room visits for respiratory complaints from the study of Glasser, Greenburg, and Field[39] together with our own unpublished data (lower panel). Thus, four separate health indices demonstrated a significant increase during the period of high pollution.

Non–episodic-correlated health effects. Health effects of community air pollution are most obvious during the episodes of unusually high pollution reviewed in the previous section. Evidence to be reviewed in this section suggests that exposure to regularly occurring concentration of air pollution may also exert a deleterious effect on health. The precautions regarding epidemiologic interpretations discussed earlier apply with equal urgency to the studies reviewed here. An observed health effect may be due to variables other than air pollution or an actual health effect due to air pollution may be masked by some more powerful factor such as ciga-

Table III. *Air quality measurements in regions used for epidemiologic study*

	High	Low	High/low ratio
Prindle et al.[80]			
Sulfation, mg. SO_3/100 cm.²/day	3.7	0.6	6.2
Suspended particulates, μg/M.³	151	109	1.39
Dustfall, tons/mile²/month	83	26	3.2
Holland and Reid[57]			
Sulfation, mg. SO_3/100 cm.²	1.65	0.95	1.7
Dustfall, mg./M.²	136	59	2.3
Burn and Pemberton[19] *(winter means)*			
Sulfur dioxide, p.p.m.	0.25	0.12	2.08
Smoke concentration, μg/M.³	770	450	1.7
Winkelstein[104]			
Sulfation, mg. SO_3/100 cm.²/day	1.25	0.20	6.3
Suspended particulates, μg/M.³	206	75	2.7
Dohan et al.[29]			
Suspended sulfate, μg/M.³	19.8	7.4	2.7
Suspended particulates, μg/M.³	173	101.	1.7
Toyama[95] *(Kawasaki City—April)*			
Sulfation, mg. SO_3/100 cm.²/day	1.7	—	—
Dustfall, tons/Km.²/month	70.0	12.2	5.7
Ishikawa et al.[60] *(emission data)*			
Sulfur oxides, tons × 10³/year	455	36	12.6
Particulates, tons × 10³/year	147	82	1.8

This table reflects the variety of air pollution indices found in the literature. Since the results from one instrument may not be directly translated to those from another, no attempt at conversion has been made.

rette smoking.

Obviously all such papers could not be included. We have in general selected papers which have either attempted to control most extraneous variables or have presented important observations that could be due to positive findings. While negative findings frequently do not find their way into print, several such papers have emphasized that "negative" findings cannot be equated with "no air pollution effect." Louden and Kilpatrick,[68] for example, were unable to find a correlation between the frequency with which antitussive medication was prescribed and the concentrations of certain air pollutants. They pointed out that this technique might not be sufficiently sensitive to demonstrate an effect of air pollution on the symptom of cough.

Geographic differences in disease frequencies. A number of studies have demonstrated an increased incidence of respiratory disease among urban populations compared with that among rural populations. Since the demonstration of an urban-rural health gradient obviously depends upon the demonstration of an urban-rural pollution gradient, pollutant ratios (for indices of sulfation and particulate pollution) are given. Absolute levels of pollution in the studies cited are presented in Table III to enable comparison of air quality among areas. Moreover, the studies discussed emphasize that other variables such as economic level, indoor heating techniques, and frequency of cigarette smoking may also produce an urban-rural health gradient.

Prindle and associates,[80] in 1959, studied two neighboring communities in Pennsylvania which were believed to have different air quality characteristics. Dustfall and sulfate pollution were 3.2 and 6.2 times

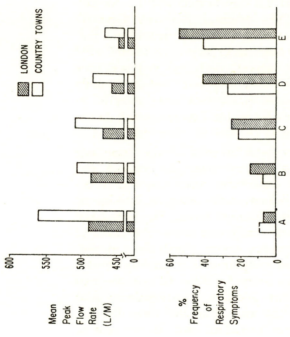

Fig. 6. The interrelationships between smoking and exposure to air pollution are shown. Data obtained from men living in London are shown in hatched area; data from men living in rural areas are shown in clear areas. Smoking history as follows: *A*, nonsmoker; *B*, exsmoker; *C*, 1 to 14 Gm. tobacco per day; *D*, 15 to 24 Gm. per day; *E*, 25 Gm. or more. Note that the frequency of cough and sputum production increases and mean peak flow rate decreases with increasing smoking exposure. For each smoking class, symptoms are more frequent and pulmonary function poorer in men living in London compared to those for men living in rural areas. (From Holland and Reid.[57])

more intense in Seward as compared with New Florence. In addition, sulfur dioxide vegetation damage was demonstrated in Seward but not in New Florence. While most pulmonary function tests were similar, subjects in Seward had significantly higher airway resistances suggesting a subtle effect of some environmental factor.

Respiratory disease incidence was studied in 293 London men and 477 rural Englishmen by Holland and Reid.[57] Pollutant ratios averaged 2.3 for dustfall and 1.7 for sulfation. The Medical Research Council's short questionnaire and several simple pulmonary function studies were administered to each subject. Both age and residence gradients were observed. Persistent cough and sputum production were observed in 38.7 per cent of the London men between the ages of 50 and 59 years compared to 18.9 per cent in men of the same age living in country towns. In contrast, significant differences were not found between younger men living in London or in English country towns. Fig. 6 is taken from their study and shows the effect of both cigarette smoking and place of residence on symptoms of respiratory disease and pulmonary function. Clearly the two factors are additive. The Londoner who smokes has more than five times the frequency of bronchitis and 80 per cent of the pulmonary function of a rural dweller who does not smoke.

Holland and Reid later combined their data with those obtained from American telephone workers by Seltser and Stone. The combined data[58] provide an interesting perspective on the comparative frequency of respiratory disease in the two countries. The frequency of persistent cough and sputum production was 22.2 per cent in the fifth decade and 25.8 per cent in the sixth decade in the American

119

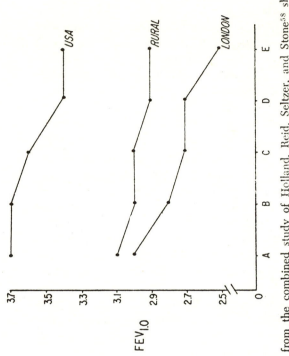

Fig. 7. Data from the combined study of Holland, Reid, Seltzer, and Stone[58] show the relationship of a single pulmonary function test, FEV₁, to smoking history and place of residence. Smoking classes on abcissa are the same as in Fig. 6.

subjects. These combined data suggest that little difference exists in the frequency of chronic bronchitis among men between the ages of 40 and 49 living in London, English country towns, or the United States. In the decade 50 to 59, however, bronchitis is much more common in London than in either English country towns or the United States. A marked age gradient is seen in the London men but not in the other two groups suggesting that life in that city predisposes to the development of bronchitis. Fig. 7, taken from the combined study, shows the relationship of a single pulmonary function test, the FEV_1 (one-second forced vital capacity), and cigarette smoking in the three populations. Some factor in addition to cigarette smoking appears to depress pulmonary function in Englishmen, and this appears to be the levels of air pollution.

A third study by Colley and Holland[24] shows the application of the questionnaire technique to family groups in an effort to assess the varying influences of smoking, area of residence, place of work, family size, social status, and genetic factors in the development of chronic respiratory disease. Two areas of London were selected for study. In fathers, smoking and social class influenced the prevalence of cough while the area of home residence did not. In both children and mothers the area of residence influenced the incidence of cough; mothers were not influenced by social class. Studies such as this are useful in attempting to separate the interacting factors influencing the frequency of respiratory disease in a given population.

Burn and Pemberton[19] took advantage of differences in air pollution within the town of Salford to study the effects of air pollution on the incidence of chronic

121

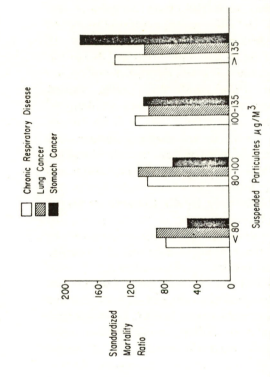
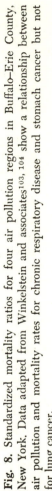

Fig. 8. Standardized mortality ratios for four air pollution regions in Buffalo–Erie County, New York. Data adapted from Winkelstein and associates[103, 104] show a relationship between air pollution and mortality rates for chronic respiratory disease and stomach cancer but not for lung cancer.

bronchitis and lung cancer. The average smoke pollution for the entire area is extremely high (510 μg per cubic meter) but ranges from 170 on the periphery to 680 centrally. Measurements of both sulfur dioxide and particulate concentrations indicate air pollution in the central areas to be about four times more intense than that in outlying areas. Incidence ratios based on observed incidence divided by expected incidence were calculated and tested by chi square analysis. Incidence ratios for episodes of bronchitis were 130 in the high pollution area and 60 in the low pollution area. Mortality ratios for deaths from bronchitis, lung cancer, arteriosclerotic heart disease, and cerebral vascular disease were 128, 124, 90, and 75, respectively, in the high pollution area and 52, 79, 120, and 111 in the low pollution area. Differences in morbidity and mortality rates from chronic bronchitis and death from lung cancer were highly significant.

A similar geographic approach within the greater Buffalo area was undertaken by Winkelstein and associates[103, 104] and demonstrated similar findings. In addition, this study also evaluated economic status as a possible causal factor in death from chronic obstructive lung disease. Standardized mortality ratios were calculated from death certificates, and economic class was derived from census information. Mortality rates from respiratory disease and gastric carcinoma within each economic group were closely related to air pollution levels which averaged greater than 135 μg per cubic meter in the high pollution area and less than 80 μg per cubic meter in the low pollution area. A similar relationship could not be established for lung cancer. Correlation between mortality rate and air pollution with the use of sulfation (lead peroxide candle method) as the index of pollution was poor, demonstrating the importance of selection of air quality indices in such epidemiologic studies.[105] Fig. 8 shows the standardized mortality ratios for chronic respiratory disease, lung cancer, and gastric carcinoma for the four air pollution areas.

Two other studies demonstrate the advantages and pitfalls of correlative studies among illness rates and pollution levels in multiple geographic areas. Hickey and associates,[56] in a startling and provocative paper, correlated atmospheric concentrations of a number of trace metals and of suspended particulates with mortality data from a number of disease states for 26 American cities. Correlation coefficients between cadmium, zinc, tin, and vanadium concentrations and diseases of the cardiovascular system were 0.73, 0.56, 0.55, and 0.50, respectively. Lung cancer was correlated with the same trace metals but not with suspended particulates. Such data must be considered as an opening epidemiologic wedge rather than evidence that cadmium causes heart disease since some unspecified variable might affect both trace metal pollutant levels and mortality rate from heart disease. Dohan and associates[29] correlated the incidence of respiratory disease absences lasting more than 7 days for five plants located in cities with varying air pollution. A close correlation between absence rates and suspended sulfates (correlation coefficient, 0.96) but not with suspended particulate was demonstrated. A similar relationship at a higher level was demonstrated when a year with an influenza epidemic was compared with other nonepidemic years.

The problem of air pollution is worldwide and similar geographic studies have appeared from other countries. Toyama[95]

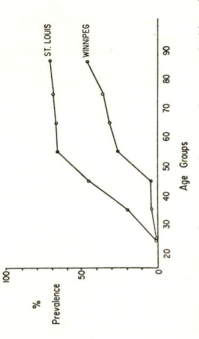

Fig. 9. Incidence of anatomic emphysema as a function of age in a highly polluted city, St. Louis, compared to a relatively clean city, Winnipeg. (Redrawn from Ishikawa and associates.[50])

recently reviewed air pollution research in Japan and demonstrated high correlation coefficients between bronchitis and air pollution in 21 districts of Tokyo but not for lung cancer or cardiovascular disease. Oshima and associates[77] administered a health questionnaire to a group of workers and found respiratory symptoms to be considerably more common in the high pollution Tokyo-Yokohama area compared to the low pollution Nigata area.

Another epidemiologic approach is the study of geographic variations in observed pathology. Thurlbeck[94] has recently reviewed the criteria for the pathologic diagnosis of emphysema and outlined an approach to "geographic pathology." Toyama[95] studied the incidence of pulmonary fibrosis in accidental deaths autopsied by the Tokyo Municipal Medical Examiner. Fibrosis was found in 49 per cent of lungs from persons living in highly polluted areas but only 16 per cent of lungs from those living in areas of low pollution.

Ishikawa and associates[60] used a quantitative technique to measure the presence and severity of emphysema in 300 routine autopsies from each of two cities with differing air quality. St. Louis, Missouri, is heavily contaminated emitting six to thirteen times the pollutants as Winnipeg in Canada. Their findings are shown in Figs. 9 and 10 adapted from their study. The incidence of emphysema increases with age in both cities but is always considerably more prevalent in St. Louis. For each category of smoking history, emphysema was always more common in St. Louis. The interacting role of cigarette smoking is seen in the observation that severe emphysema although more common in smokers living in St. Louis than in Winnipeg was never seen in nonsmokers living in either city. Such studies may explain the

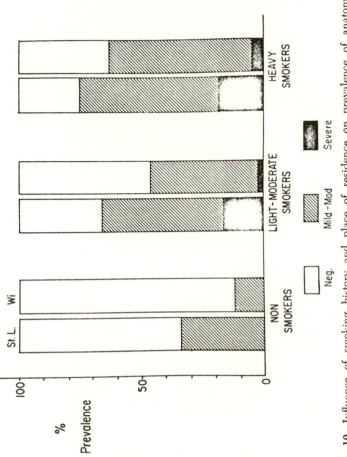

Fig. 10. Influence of smoking history and place of residence on prevalence of anatomic emphysema from study of Ishikawa and associates.[60] For each smoking history more emphysema was found in St. Louis compared to Winnipeg. Differences were less marked in heavy smokers. Note that severe emphysema was not seen in nonsmokers living in either city.

wide variation in incidence of emphysema which ranges from 12 per cent in the nonsmoking resident of Winnipeg to almost 80 per cent in the smoker from St. Louis.

A number of informative epidemiologic studies have been performed on children since variation due to tobacco smoking, occupation, and frequent change of address is generally minimized. Douglas and Waller[31] studied a sample of 3,866 children from birth in 1946 to the age of 15 in 1961. Four air pollution areas, based on domestic coal consumption, were constructed. Table IV lists several of the many indices reported in this study and reveals a consistent relationship between lower (but not upper) respiratory disease and level of air pollution.

Toyama[95] studied pulmonary function and administered a questionnaire to a large group of children attending six schools in the Tokyo area. Pollution levels were about three to four times higher in schools located in industrial areas compared to those in residential areas. The frequency of cough and eye irritation ranged from 2.0 and 13.8 per cent, respectively, in the lowly polluted area to 13.2 and 59.8 per cent in the high pollution area, while peak flow rate measured during February averaged 306 ± 3.7 L. per minute in the low pollution area and 270 ± 3.3 in the high.

Temporal variations in mortality rate and morbidity. In addition to studying geographic variations in illness rates, the air pollution epidemiologist may attempt to study temporal variations in disease rates within a single community and attempt to correlate these variations with certain environmental variables.

One of the earliest demonstrations of the relationship between illness rates and air pollution levels was the study of Waller

Table IV. *Incidence of respiratory infections in children (per cent of sample) from Douglas and Waller[31]*

	Air pollution exposure			
	Very low	Low	Moderate	High
Hospital admission for:				
Bronchitis	0.0	0.9	1.0	1.4
Pneumonia	1.1	1.4	1.6	1.8
Upper respiratory infection	0.4	0.3	0.4	1.1
Tonsillitis	4.4	6.2	5.7	5.2
More than one attack of lower respiratory infection in first two years of life				
Middle class	3.0	4.0	7.7	9.3
Working class	5.1	10.8	13.9	15.4
Rales or rhonchi recorded				
Once or more	10.8	13.7	17.5	17.3
Twice or more	0.5	2.1	2.5	2.7
At age 15	0.2	1.2	2.1	2.2

and Lawther[99] who followed 180 bronchitic patients in Greater London by means of regular diary entries. The patients entered a single letter in their diary each day and this was scored as follows: condition better than usual, –1; condition unchanged, 0; condition worse than usual, +1; condition much worse than usual, +2. Smoke and sulfur dioxide concentrations were measured by several techniques. Fig. 11 reveals the striking correlation in smoke and illness peaks during a 4 month period in 1955 to 1956. A smog episode began during the first week of January and smoke concentration reached the extremely high level of 10 mg. per cubic millimeter later in the evening of January 3. Early the next morning, a wet fog appeared decreasing visibility and smoke concentration fell to almost normal limits. The illness score exceeded 0.7 indicating that, on the average, 70 per cent of the chronic bronchitics felt worse. The authors concluded that air pollution appeared to exert significant adverse health effects but that dense wet fog did not appear to intensify these effects.

A detailed study of "Health and the urban environment" was undertaken by McCarroll and associates[70] within a half-mile square area in the lower east side of Manhattan. A daily record of such symptoms as cough and burning or itching eyes was kept in over 1,000 adults for a full year. Smoking history was recorded and air pollution measured by a specially constructed monitoring station within the area.

Their data were initially analyzed by a series of correlation coefficients between each symptom and each measure of air pollution. These cross-correlations were calculated with a time lag of between 0 to 28 days since the appearance of a symp-

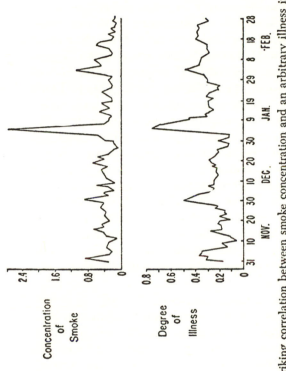

Fig. 11. Striking correlation between smoke concentration and an arbitrary illness index in 180 bronchitics studied by Waller and Lawther[60] in Greater London during the winter of 1955 to 1956. This study provided the first experimental evidence that the health of patients with chronic obstructive lung disease was impaired by exposure to atmospheric pollutants.

tom might be related to environmental events occurring at some point in the past. In addition, autocorrelations, correlations between each symptom, and index of pollution and the corresponding value for each preceding day back to 28 days were calculated as indices of phenomena persistence. For rapid inspection purposes, each set of correlations was plotted against the lag in days as multiple correlograms.

The data demonstrated that eye irritation lasted for short periods of time while cough persisted for considerably longer periods of time. The best correlations between eye irritation and sulfur dioxide appeared when lag was not introduced; the best correlation between cough and sulfur dioxide appeared when levels of pollutant measured one to three days prior to onset of cough were used. A similar relationship between cough and particulate density was observed; eye irritation did not correlate with particulate density.

The same data were reevaluated by Cassell and associates[21] in a recent paper. Noting that previous papers had pointed to a multivariate nature of both stimulus and response, the authors report results from multiple correlation analysis and from a relatively new statistical technique, principal components analysis. Correlation coefficients between the symptom of cough and wind speed were 0.217; temperature, –0.494; barometric pressure, 0.269; particulate pollution, 0.194; and sulfate pollution was 0.112.

Principal components analysis was undertaken to study the interaction of groups of environmental factors on respiratory symptoms. The symptom complex of cough, sore throat, and "common cold" was found to be associated with two sets of environmental conditions: cold, wet, windy, "terrible" weather or cloudy, humid, not too cold, windless weather associated with atmospheric stagnation. Eye irritation, in contrast, was associated with sunny, windless days during periods of high automotive pollution.

The multiple-factor problem may be well seen in the studies of Spodnik and associates[89] and of Verma and associates,[98] who serially studied airway resistance and absence rates, respectively. Both groups noted a strong cyclical time trend. Airway resistance was observed by Spodnik and associates to be closely correlated with outdoor temperature (−0.46) and only poorly associated with particulate concentration. Unfortunately, only a single index of air pollution, suspended particulate concentration, was included in the experimental design. Verma and associates found that 46 per cent of the respiratory illness rate variability was accounted for by a time model and 20 per cent of the variability by a linear air pollution model, while combination of both models explained 50 per cent of the variation. They concluded that respiratory illness rates were highest on cool days when sulfur dioxide levels exceeded 0.05 p.p.m.

An informative study on the appearance of minor respiratory illnesses in factory and office workers was undertaken by Angel and associates[9] from the Hammersmith Hospital. A questionnaire, pulmonary function studies, and sputum collections were made every 3 weeks on a sample of 92 workers. The classification and frequency of respiratory illness observed in the period between October, 1962 and May, 1963 were: simple coryza, 53; chest colds, 57; mucopurulent bronchitis, 39; influenza-like illnesses with systemic symptoms, 29; and wheezy attacks, 21. That mysterious "pathogen" *Hemophilus influenzae* was found in 9 per cent of specimens between illnesses but in 25 per cent of specimens with mucopurulent bronchitis. Correlation coefficients between the attack rate of respiratory illness and smoke were 0.57, sulfur dioxide was 0.56, and temperature was –0.26. The first two were significant at the one per cent level.

An extensive study of hospital admissions for certain "relevant diseases" based on Blue Cross Hospitalization Plan Data was undertaken by Sterling and associates.[90] The large number of observations rendered even small correlation coefficients strikingly significant. Correlations were markedly improved by the addition of nonlinear and multiple variable analysis and demonstrated a close relationship between a number of environmental variables and admission rates. The multiple correlation coefficient including linear and nonlinear factors and interactions for pollutant concentrations, temperature, and humidity was 0.373 with 4,564 degrees of freedom.

Ipsen and associates[59] studied industrial absenteeism from upper respiratory infections and observed the illness rate to correlate with temperature (–0.614), particulate concentration (0.556), and particulate sulfate concentration (0.289).

The studies reviewed in this section reveal a strong time dependence of respiratory infection and other illnesses. Interrelationships among individual pollutants and other environmental factors such as temperature, humidity, and barometric pressure complicate interpretations and force reliance on complicated statistical techniques. The data suggest, however, a meaningful relationship between urban air pollution and health.

Asthma and air pollution. Episodic bronchospasm has dramatically increased in urban areas suggesting the causal role of infectious and pollutant agents rather than classical pollen allergy. Certain outbreaks of bronchospasm have assumed sufficient proportion to be assigned generic names. Booth and associates[16] from the United States Public Health Service studied emergency room visits for asthma in ten hospitals located in seven cities and found great similarity in seasonal peaks with major increases in asthma visits occurring in the fall months after the completion of the pollinating season. Peaks during months of high pollen count were not seen. These observations have stimulated a search for possible environmental factors on the development of episodic bronchospasm.

The striking increase in asthma admissions in New York City was revealed in a study by Greenburg and associates.[44] In 1952, 3.2 and 6.6 per cent of emergency room visits at Harlem and Metropolitan Hospital were for asthma. By 1962, these percentages had risen to 25.7 and 15.7 per cent. An attempt to correlate asthma attacks with environmental factors was made by Girsh and associates[38] in Philadelphia who found a threefold increase in asthma on days of high air pollution, a fourfold increase during days of high barometric pressure, and a ninefold increase during periods of stable atmosphere when both high barometric pressure and high pollutant levels coincided.

A unique form of asthma, Tokyo-Yokohama asthma, was first observed at the United States Army Hospital in Yokohama in 1946.[79] Large numbers of American servicemen became afflicted with a condition characterized by nocturnal coughing and wheezing which disappeared when they left the highly polluted Kanto Plain region of Japan. The disease is more properly termed bronchitis than asthma, is more common in smokers, and appears to be related to both irritant and allergenic components of the ambient air. Another outbreak of asthma, New Orleans asthma, appears to predominantly affect allergic individuals without prior evidence of bronchitis. Patients with this syndrome had a significantly higher rate of positive skin tests to extracts of atmospheric pollutants than did control subjects without bronchospasm.[101]

Carbon monoxide

The toxicity of carbon monoxide has been the subject of many reviews since the series of courageous experiments performed upon himself was first reported in 1895 by John Haldane.[54] Haldane did not observe serious toxic symptoms until at least one third of his blood was saturated with carbon monoxide, and most subsequent investigations emphasized dramatic toxic effects of exposure to high doses. Important as these early investigations were, they tended to minimize the effects of exposure to low dosages and tacitly implied that carboxyhemoglobin levels associated with cigarette smoking and urban air pollution were probably harmless. In this review we propose to summarize recent work pointing toward important physiologic effects of low concentrations of carbon monoxide as well as present current concepts regarding mechanisms of action.

In many respects, carbon monoxide serves as a model pollutant. Ideally, one would like to identify the body burden of a given pollutant and be in a position to measure the physiologic effects of that particular body burden. With most pollutants, neither the body burden nor the quantitative relationships between burden and health effects are known. With carbon

136

monoxide, in contrast, one can estimate body burden from blood levels and determine quantitative relationships by experimental observations made during the administration of known concentrations.

Basic chemical reactions. Both oxygen and carbon monoxide combine reversibly with hemoglobin to form oxyhemoglobin and carboxyhemoglobin, respectively. Douglas, Haldane, and Haldane[30] first demonstrated that the toxicity of extremely low concentrations of carbon monoxide was due to its great affinity for hemoglobin relative to oxygen since the partial pressure of carbon monoxide required to fully saturate hemoglobin is only $\frac{1}{200}$ to $\frac{1}{250}$ of the partial pressure of oxygen required for complete saturation with oxygen. When exposed to a mixture of both gases, equilibrium is gradually reached according to the Haldane equation:

$$\frac{HbCO}{HbO_2} = M\,\frac{P_{CO}}{P_{O_2}}$$

where $HbCO$, HbO_2, P_{CO} and P_{O_2} represent hemoglobin saturations and partial pressures of carbon monoxide and oxygen, respectively. M is a constant expressing relative affinities which ranges from 200 to 250 in human blood. At equilibrium, for example, environments containing 25, 50, and 100 p.p.m. of carbon monoxide would lead to carboxyhemoglobin saturations of 4.7, 9.1, and 16.1 per cent if arterial oxygen tension were 80 mm. Hg.

The time required to reach equilibrium varies widely and is related to minute ventilation, dead space volume, diffusion capacity of the lung, total red cell mass, rate of blood flow, and COHb concentration in venous blood. Forbes and associates,[34] Pace and associates,[78] and Lilienthal and Pine[67] have studied the time course

[Handwritten margin note: Low levels of CO bind hemoglobin with an affinity 100 to 200 times that of O2. Myoglobin also binds CO more avidly than oxygen.]

137

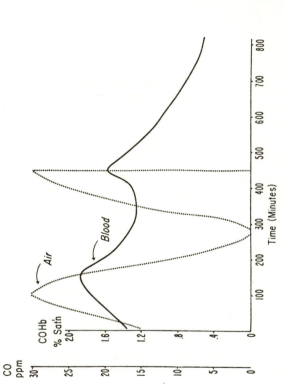

Fig. 12. The relationship between ambient air concentrations of carbon monoxide and blood carboxyhemoglobin levels. Note that blood levels change much more slowly than ambient air concentrations and are a damped representation of events occurring in the atmosphere. For this reason, measurement of blood levels provides a better estimate of the body burden of carbon monoxide. (Redrawn from Goldsmith and associates.[43])

of carboxyhemoglobin during the steady-state breathing of various mixtures of carbon monoxide. While the relationships are complex and frequently exponential, the latter group found good agreement with experimental data with the use of a linear equation: COHb = (Pco in inspired air) (exposure time in minutes) (ventilatory rate in liters per minute) (0.05). Solution of this equation reveals that carboxyhemoglobin saturation in a lightly exercising man in an environment containing 50 p.p.m. of carbon monoxide would be 1.4, 2.3, 3.4, 4.6, and 5.7 per cent saturation at 1, 2, 3, 4, and 5 hours respectively.

While the physiologic variables listed above confuse the prediction of carboxyhemoglobin concentrations in a given individual, the obvious fact that individuals do not breathe a constant concentration of carbon monoxide further interferes with attempts at predicting blood levels from ambient air concentration. An individual may breathe 5 p.p.m. for several hours, spend 15 minutes driving through a tunnel containing 150 p.p.m., and then remain in traffic for 30 minutes breathing an atmosphere ranging from 20 to 50 p.p.m.

These relationships would be even further confused if he smoked several cigarettes, since cigarette smoke contains 3 to 4 per cent carbon monoxide and each diluted puff probably is equivalent to about 500 p.p.m. of carbon monoxide.

Goldsmith and associates[43] have attempted to study this problem of fluctuating exposures by developing a computer program which could rapidly calculate carboxyhemoglobin saturations from certain assumed constants and a changing environmental carbon monoxide concentration. Fig. 12 is redrawn from their work and shows that the blood concentration is a damped representation of the air concentration with a definite lag between the two systems.

Assessment of human exposures to carbon monoxide. The human exposure to carbon monoxide, like that to other pollutants, may be estimated by the frequent measurement of atmospheric concentrations. While the concentrations of many other pollutants are relatively consistent within a given geographic area, carbon monoxide concentrations may vary widely because of the multiple point sources responsible for its production. A sensing probe at bumper level in traffic may reveal an extremely high level which progressively falls off as samples are taken at driver level, on the sidewalk, or on the second and third floors.

Average carbon monoxide levels taken from probes located in laboratories somewhat remote from traffic may vary from 2 to 50 p.p.m. In many urban areas, these levels are consistently below 10 p.p.m. but may have little relationship to events in traffic-congested areas. Ramsey[51] studied the carbon monoxide concentration at curbside in a number of traffic intersections in Dayton, Ohio. Samples were taken at 5 feet and averaged 36 p.p.m. with a range from 2 to 135 p.p.m. Haagen-Smit[50] equipped an automobile with sensing equipment and detected peak values of 120 p.p.m. in Los Angeles traffic. Miranda and associates[72] recently reviewed data from several tunnels in an effort to determine tunnel control techniques. Carbon monoxide concentration averaged 65 p.p.m. but peaked to 390 p.p.m. during a period of severe traffic congestion in the Holland Tunnel. Concentrations on the expressway leading to the George Washington Bridge in New York City were found to average 40 p.p.m. for the month, although peaks of from 200 to 300 p.p.m. were occasionally encountered.

The theoretic considerations outlined above suggest that measurements of blood carbon monoxide concentrations provide a more consistent index of human exposure than measurement of ambient air concentrations. Fortunately, a number of extremely sensitive blood methods exist but the need for venesection limits the widespread epidemiologic application of this technique. ⊙

Alveolar and blood carbon monoxide tensions are quite similar permitting the use of alveolar tensions to estimate blood levels. Jones and associates[62] developed a breath-holding method which was sensitive enough to show that 90 per cent of the values obtained fell within ± 1.3 per cent saturation of the mean curve, more than adequate for epidemiologic studies. While direct blood measurements give the most reliable estimates of human exposure to carbon monoxide, alveolar sampling techniques appear to be almost as accurate and considerably more relevant than measurement of atmospheric levels.

Kjeldsen[63] recently reviewed reported observations on blood carboxyhemoglobin. The average carboxyhemoglobin concentration in 511 nonsmokers was 1.0 and ranged from 0 to 4.0 per cent saturation. The average carboxyhemoglobin in 509 smokers was 3.9 and ranged to levels as high as 19 per cent saturation. Fig. 13 shows the distribution of blood carboxyhemoglobin concentrations (determined by alveolar air sampling) in a series of observations made in a congestive area in New York City in cooperation with Dr. Eric Cassell and Commissioner Austin Heller. The distribution of concentrations was similar for the two groups in the lower range of concentrations but a splayed-out

141

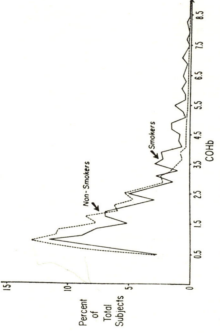

Fig. 13. Distribution of carboxyhemoglobin levels in 1,481 subjects studied in Herald Square, New York City, by the authors in cooperation with Dr. Eric Cassell and Commissioner Austin Heller. The dotted line represents nonsmokers, the solid line, smokers. Ordinal values are percentage of entire sample. Alveolar sampling techniques were used to estimate blood carboxyhemoglobin. The distributions are similar for the two groups, although smokers show a splayed-out tail with concentrations up to 14 per cent carboxyhemoglobin.

tail of the distribution curve revealed that a small number of smokers had carboxyhemoglobin concentrations up to 14 per cent saturation. These studies were performed during a period of relatively low air pollution.

Physiologic effects of carboxyhemoglobin. The physiologic effects of carbon monoxide toxicity resemble the effects of hypoxia, and most investigators have considered that the primary action of carbon monoxide is exerted through its competition with oxygen for binding sites on the hemoglobin molecule. While undoubtably this is the major site of action, certain recent observations suggest that carbon monoxide may also have a direct effect on other iron-containing compounds (possible cytochrome oxidase or cytochrome a_3) or other yet unknown sites.[43]

Douglas and associates[30] showed that the dissociation curves for either carbon monoxide or oxygen with hemoglobin are identical, provided that the abscissal values for carbon monoxide pressure are multiplied by the affinity constant, M. In contrast, the oxyhemoglobin dissociation curve in the presence of COHb and reduced hemoglobin is unique and not classically sigmoid, but shifted to the left so that a lower oxygen tension exists for the same oxyhemoglobin saturation compared to blood without COHb present (Fig. 14). These investigators pointed out that this characteristic shift of the oxyhemoglobin dissociation curve explained the "contrast between the helpless condition of a person whose blood is half saturated with carbon monoxide and the comparatively slight symptoms when the hemoglobin is reduced to half its normal percentage in anemia." Carbon monoxide not only diminishes the total amount of oxygen available (by direct replacement of oxygen) but also alters the dissociation of the remaining oxygen so that it is held more tenaciously by hemoglobin and released only at lower oxygen tensions.

The oxyhemoglobin curve in the presence of COHb progressively resembles that of myoglobin as the concentration of COHb is increased (Fig. 14). Since myoglobin is a heme compound with only one heme unit per molecule and does not exhibit heme-heme interactions, it is possible that the combination of one or more of the four heme groups in hemoglobin with carbon monoxide decreases the heme-heme interactions of the remaining heme units and results in a molecule approaching the behavior of myoglobin.

The impact of conversion of part of the circulating oxyhemoglobin to carboxyhemoglobin may be seen in experimental studies performed during routine cardiac catheterization. We[13] recently acutely raised carboxyhemoglobin concentration to an average of 9.0 per cent saturation in 26 subjects with or without heart disease. Mixed venous oxygen tension decreased from 39 to 31 mm. Hg suggesting a significant decrease in tissue oxygen tension since mixed venous oxygen tension must represent a maximum value for tissue oxygenation. Surprisingly, arterial oxygen tension decreased from 81 to 76 mm. Hg. Early investigators had suggested that arterial oxygen tension was unchanged during carboxyhemoglobin changes. Measurement of the alveolar-arterial oxygen difference revealed an increase from 20 to 29 mm. Hg and suggested that an augmentation of the shunt effect was probably responsible for both the increase in alveolar-arterial oxygen difference and decrease in oxygen tension. Brody and Coburn[18] have recently confirmed these findings.

Effects of low concentrations of COHb

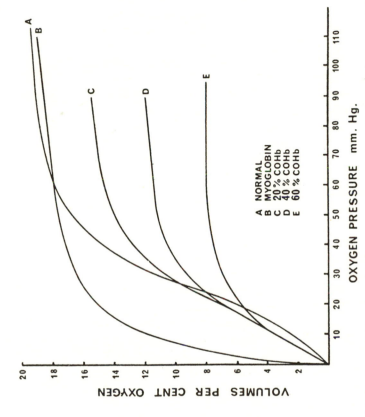

Fig. 14. Oxyhemoglobin dissociation curves for normal adult hemoglobin, myoglobin, and varying mixtures of oxyhemoglobin and carboxyhemoglobin. The presence of significant amounts of carboxyhemoglobin (curves *C*, *D*, and *E*) reduces the total amount of oxygen carried and also changes the shape of the curve so that a lower oxygen tension results from the same oxygen content. At lower oxygen tensions, the carboxy-oxyhemoglobin curves resemble the dissociation curves for human myoglobin.

on the central nervous systems. A number of recent studies have presented objective evidence that relatively low levels of COHb can interfere with certain of the higher integrative functions of the central nervous system. MacFarland and associates[71] were able to show a linear relationship between decrease in visual discrimination and increase in COHb and observed significant impairment in discrimination with COHb levels as low as four per cent saturation. Lilienthal and Fugitt[66] observed an impairment in flicker-fusion with COHb levels between 5 and 10 per cent at 6,000 feet altitude, and Trouton and Eysenck[96] demonstrated difficulty in limb coordination at similar concentrations of COHb. A battery of physiologic and psychologic tests was applied to a group of normal subjects by Schulte[55] before and after they breathed sufficient carbon monoxide to raise COHb levels to between 0 and 20.4 per cent. Although significant correlations between pulse rate, blood pressure or respiratory rate and COHb were not observed, Schulte was able to closely relate seven psychologic responses to COHb. COHb concentrations below three per cent produced significant increases in errors in letter and color response tests, in the completion time of plural noun underlining tests, and other similar psychomotor tests. Similar findings were reported by Beard and Wertheim who observed psychologic impairment at levels below two to three per cent carboxyhemoglobin.[14] Such psychologic studies confirm the neurophysiologic observation of Xinteras and associates[107] who observed alteration in evoked potentials from the superior colliculus of unrestrained and unanesthetized rats following the inhalation of low concentrations of carbon monoxide.

Cardiovascular effects. A number of

studies on the hemodynamic effects of carbon monoxide inhalation have been described. Both Haldane[54] and Haggard[53] reported hyperventilation in man and in experimental animals. The contrast between hypoxemia caused by breathing low concentrations of oxygen and carbon monoxide poisoning was stressed by Asmussen and Chiodi[10] who observed a marked increase in both minute ventilation and cardiac output when arterial oxygen tension was lowered but not when carboxyhemoglobin levels were raised to 20 to 30 per cent. Although carboxyhemoglobin did not affect cardiac output, it increased pulse rate substantially leading the authors to suggest a compensatory chronotropic response to a primary decrease in stroke output. In a later more detailed study, Chiodi and associates[23] observed cardiac output and pulse rate to rise together with increasing concentrations of carboxyhemoglobin. While relatively high concentrations were evaluated, changes in pulse rate and cardiac output were noted with as little as 16 per cent carboxyhemoglobin.

We recently reported a series of systemic and coronary hemodynamic studies in 26 human beings and made additional measurements in a group of animals studied at high concentrations of carboxyhemoglobin.[13] Raising carboxyhemoglobin saturation to an average of nine per cent increased minute ventilation, cardiac output, and the oxygen extraction ratio. Similar concentrations increased coronary blood flow but decreased myocardial oxygen extraction so that myocardial oxygen consumption was essentially unchanged. Changes in lactate and pyruvate extractions suggesting anaerobic metabolism were observed in certain individuals with coronary artery disease and in one patient with obstructive emphysema and hypoxemia.

The effects of rest and exercise on oxygen uptake and heart rate before and after elevation of carboxyhemoglobin concentrations to four per cent were studied by Chevalier and associates.[22] This level of carboxyhemoglobin produced a significant increase in the oxygen debt of exercise and also decreased both resting pulse rates and the chronotropic response to exercise.

Many investigators have examined the effect of carbon monoxide on the electrocardiogram in patients who have been exposed to high concentrations of the gas. Cosby and Bergeron[27] studied the electrocardiogram of ten patients admitted for carbon monoxide poisoning and observed abnormalities such as sinus tachycardia, T wave abnormalities, S-T segment depression, and atrial fibrillation in nine. Five of 6 patients studied by Anderson and associates[8] had significant electrocardiographic abnormalities. A remarkable case of a young man who developed a chronic illness resembling primary myocardial disease following carbon monoxide poisoning was reported by Shafer and associates.[86] Animal studies performed by Ehrich and associates[32] demonstrated electrocardiographic abnormalities when the carboxyhemoglobin level was raised to about 20 per cent for two weeks.

Two studies from Astrup's laboratory[12, 100] demonstrated that relatively low concentrations of carboxyhemoglobin (11 per cent) produced degenerative cardiovascular changes in rabbits and also enhanced aortic atheromatosis in rabbits fed cholesterol. Kjeldsen,[63] in Astrup's laboratory, reviewed much of the data relating smoking and carbon monoxide inhalation to atherosclerosis and presented a series of original investigations designed to test the hypothesis that carbon monoxide did increase the proclivity of experimental animals to develop atherosclerosis. Twenty-four rabbits were fed cholesterol and the carboxyhemoglobin concentration raised to between 10 and 20 per cent in twelve. Three fourths of the experimental animals developed focal degenerative changes with atheromatosis, while similar changes were observed in only one control animal. In another study, similar carboxyhemoglobin concentrations were observed to increase serum cholesterol in animals fed both standard and cholesterol-enriched diets. While similar effects have been observed with hypoxia, these studies suggest an additional mechanism for the relationship between carbon monoxide and heart disease.

The experimental studies performed in Astrup's[11] laboratory were suggested by clinical observations demonstrating that an abnormality in the oxyhemoglobin dissociation curve was responsible for ischemic symptoms in Buerger's disease and non-specific myocarditis. Later work convinced Astrup that carboxyhemoglobin might well be the cause of the hemoglobin abnormality. More recently, Elliot and Mizukami[33] demonstrated that the blood of smokers with angina pectoris contained hemoglobin with an abnormal oxyhemoglobin dissociation curve. These patients, as well as Astrup's patients, may have abnormally high carboxyhemoglobin levels which could be responsible for their ischemic symptoms.

The experimental and clinical data presented above suggest that carboxyhemoglobin levels between 5 and 10 per cent (and in some situations substantially lower) can produce significant neural and cardiovascular effects. Such effects might be predicted to be particularly important in patients with pre-existing vascular disease.

Conclusion: Toward a healthier environment

The data discussed in this review lead to the inescapable conclusion that community air pollution is capable of producing serious health effects. The Clean Air Act of 1967 directed the Surgeon General to develop "air quality criteria" which would be implemented by the several states in the form of air quality standards. Although much of the scientific work performed over the past three decades was not intended for the construction of legal criteria and standards, review and extrapolation of a massive amount of material has resulted in publication of *Air quality criteria for particulates and air quality criteria for sulfur oxides.*[74, 75]

The criteria conclude that a number of studies have demonstrated increased morbidity and mortality rates when particulate contamination exceeds 80 to 100 μg per cubic meter and sulfur dioxide concentrations exceed 0.10 to 0.20 p.p.m. These figures are annual averages and may have little relationship to the health effects of dramatic pollutant peaks. The responsibility is left to each state and urban community to develop workable standards which will reduce pollutant concentrations to safe levels.

The fields of environmental medicine—of man's adjustment to his surroundings—provide a noble example of a partnership between science and government designed to better human existence. Often beset by legislative inertia, scientific disagreement, and the reactionary interference of vested interests, the hope exists today that the clock may be rolled back and a healthful environment recreated. The critical reader of this review will recognize the many weak spots in the investigative fabric relating air pollution and human health. Hopefuly it will stimulate further investigation in these areas.

The authors thank Commissioner Austin N. Heller, New York City Department of Air Pollution Control, for his general guidance and for free access to data generated by the New York City monitoring system. We also thank Dr. Stanley Giannelli, Jr., for his close cooperation and support and Mr. John Forrest for his clerical help in preparing this manuscript.

References

1. Albert, R. E., Lippmann, M., Spiegelman, J., Liuzzi, A., and Nelson, N.: The deposition and clearance of radioactive particles in the human lung, Arch. Environ. Health 14:10-15, 1967.
2. Amdur, M. O.: The effect of aerosols on the response to irritant gases. Inhaled particles and vapours, Oxford, 1961, Pergamon Press, Inc., pp. 281-292.
3. Amdur, M. O.: The effect of high flow-resistance on the response of guinea pigs to irritants, Amer. Indust. Hyg. Ass. J. 25: 564-568, 1964.
4. Amdur, M. O.: Respiratory absorption data and SO_2 dose-response curves, Arch. Environ. Health 12:729-732, 1966.
5. Amdur, M. O., and Corn, M.: The irritant potency of zinc ammonium sulfate of different particle sizes, Amer. Indust. Hyg. Ass. J. 24:326-333, 1963.
6. Amdur, M. O., and Mead, J.: Mechanics of respiration in unanesthetized guinea pigs, Amer. J. Physiol. 192:364-368, 1958.
7. Amdur, M. O., Melvin, W. W., and Drinker, P.: Effects of inhalation of sulphur dioxide by man, Lancet 2:758-759, 1953.
8. Anderson, R. F., Allensworth, D. C., and DeGroot, W. J.: Myocardial toxicity from carbon monoxide poisoning, Ann. Intern. Med. 67:1172-1182, 1967.
9. Angel, J. H., Fletcher, C. M., Hill, I. D., and Tinker, C. M.: Respiratory illness in factory and office workers: A study of minor respiratory illnesses in relation to changes in

ventilatory capacity, sputum characteristics, and atmospheric pollution, Brit. J. Dis. Chest 59:66-80, 1965.

10. Asmussen, E., and Chiodi, H.: The effect of hypoxemia on ventilation and circulation in man, Amer. J. Physiol. 132:426-436, 1941.

11. Astrup, P.: An abnormality in the oxygen-dissociation curve of blood from patients with Buerger's disease and patients with non-specific myocarditis, Lancet 2:1152-1154, 1964.

12. Astrup, P., Kjeldsen, K., and Wanstrup, J.: Enhancing influence of carbon monoxide on the development of atheromatosis in choles-terol-fed rabbits, J. Atheroscler. Res. 7:343-354, 1967.

13. Ayres, S. M., Mueller, H. S., Gregory, J. J., Giannelli, S., Jr., and Penny, J. L.: Systemic and myocardial hemodynamic responses to relatively small concentrations of carboxy-hemoglobin (COHb), Arch. Environ. Health 18:699-709, 1969.

14. Beard, R. R., and Wertheim, G.: Behavioral impairment associated with small doses of carbon monoxide. Presented at the 94th an-nual meeting of the American Public Health Association, San Francisco, Nov. 1, 1966.

15. Becker, W. H., Schilling, F. J., and Verma, M. P.: The effect on health of the 1966 eastern seaboard air pollution episode, Arch. Environ. Health 16:414-419, 1968.

16. Booth, S., DeGroot, I., Markush, R., and Horton, R. J. M.: Detection of asthma epi-demics in seven cities, Arch. Environ. Health 10:152-155, 1965.

17. Boren, H. G.: Carbon as a carrier mechanism for irritant gases, Arch. Environ. Health 8:119-124, 1964.

18. Brody, J. S., and Coburn, R. F.: Carbon monoxide-induced arterial hypoxemia, Sci-ence 164:1297-1298, 1969.

19. Burn, J. L., and Pemberton, J.: Air pollution, bronchitis and lung cancer in Salford, Int. J. Air Water Pollut. 7:5-16, 1963.

20. Burton, G. G., Corn, M., Gee, J. B. L., Vasallo, C., and Thomas, A. P.: Response of healthy men to inhaled low concentrations of gas-aerosol mixtures, Arch. Environ. Health 18:681-692, 1969.

21. Cassell, E. J., Lebowitz, M.D., Mountain, I. M., Lee, H. T., Thompson, D. J., Wolter, D. W., and McCarroll, J. R.: Air pollution,

weather, and illness in a New York population, Arch. Environ. Health 18:523-530, 1969.

22. Chevalier, R. B., Krumholz, R. A., and Ross, J. C.: Reaction of nonsmokers to carbon monoxide inhalation, J.A.M.A. 198: 1061-1064, 1966.

23. Chiodi, H., Dill, D. B., Consolazio, F., and Horvath, S. M.: Respiratory and circulatory responses to acute carbon monoxide poisoning, Amer. J. Physiol. 134:683-693, 1941.

24. Colley, J. R. T., and Holland, W. W.: Social and environmental factors in respiratory disease, Arch. Environ. Health 14:157-161, 1967.

25. Corn, M., and Burton, G.: The irritant potential of pollutants in the atmosphere, Arch. Environ. Health 14:54-61, 1967.

26. Corn, M., and DeMaio, L.: Particulate sulfates in Pittsburgh air, J. Air Pollut. Contr. Ass. 15:26-30, 1965.

27. Cosby, R. S. and Bergeron, M.: Electrocardiographic changes in carbon monoxide poisoning, Amer. J. Cardiology 11:93-96, 1963.

28. Davidson, J. T., Lillington, G. A., Haydon, G. B., and Wasserman, K.: Physiologic changes in the lungs of rabbits continuously exposed to nitrogen dioxide, Amer. Rev. Resp. Dis. 95:790-796, 1967.

29. Dohan, F. C., Everts, G. S., and Smith, R.: Variations in air pollution and the incidence of respiratory disease, J. Air Pollut. Contr. Ass. 12:418-436, 1962.

30. Douglas, C. G., Haldane, J. S., and Haldane, J. B. S.: The laws of combination of hemoglobin with carbon monoxide and oxygen, J. Physiol. 44:275-304, 1912.

31. Douglas, J. W. B., and Waller, R. E.: Air pollution and respiratory infection in children, Brit. J. Prev. Soc. Med. 20:1-8, 1966.

32. Ehrich, W. E., Bellet, S., and Lewey, F. H.: Cardiac changes from CO poisoning, Amer. J. Med. Sci. 208:511-523, 1944.

33. Elliot, R. S., and Mizukami, H.: Oxygen affinity of hemoglobin in persons with acute myocardial infarction and in smokers, Circulation 34:331-336, 1966.

34. Forbes, W. H., Sargent, F., and Roughton, F. J. W.: The rate of carbon monoxide uptake by normal men, Amer. J. Physiol. 143: 594-608, 1945.

153

35. Frank, N. R., Amdur, M. O., and Whittenberger, J. L.: A comparison of the acute effects of SO_2 administered alone or in combination with NaCl particles on the respiratory mechanics of healthy adults, Int. J. Air Water Pollut. 8:125-133, 1964.

36. Frank, N. R., Amdur, M. O., Worcester, J., and Whittenberger, J. L.: Effects of acute controlled exposure to SO_2 on respiratory mechanics in healthy male adults, J. Appl. Physiol. 17:252-258, 1962.

37. Freeman, G., Crane, S. C., Stephens, R. J., and Furiosi, N. J.: The subacute nitrogen dioxide-induced lesion of the rat lung, Arch. Environ. Health 18:609-612, 1969.

38. Girsh, L. S., Shubin, E., Dick, C., and Schulaner, F. A.: A study on the epidemiology of asthma in children in Philadelphia: The relation of weather and air pollution to peak incidence of asthmatic attacks, J. Allerg. 39: 347-357, 1967.

39. Glasser, M., Greenburg, L., and Field, F.: Mortality and morbidity during a period of high levels of air pollution, Arch. Environ. Health 15:684-694, 1967.

40. Goldfield, E. D., Chief, Statistical Reports Division, United States Department of Commerce: Statistical Abstract of the United States, 1967.

41. Goldsmith, J. R.: Effects of air pollution on human health, in Air pollution, vol. 1, New York, 1968, Academic Press, Inc., chap. 14, pp. 547-615.

42. Goldsmith, J. R. Air pollution epidemiology: A wicked problem, an informational maze, and a professional responsibility, Arch. Environ. Health 18:516-522, 1969.

43. Goldsmith, J. R., Terzaghi, J., and Hackney, J. D.: Evaluation of fluctuating carbon monoxide exposures, Arch. Environ. Health 7:647-663, 1963.

44. Greenburg, L., Erhardt, C. L., Field, F., and Reed, J. I.: Air pollution incidents and morbidity studies, Arch. Environ. Health 10: 351-356, 1965.

45. Greenburg, L., Erhardt, C. L., Field, F., Reed, J. I., and Seriff, N. S.: Intermittent air pollution episode in New York City, 1962, Public Health Rep. 78:1061-1064, 1963.

46. Greenburg, L., Field, F., Erhardt, C. L., Glasser, M., and Reed, J. I.: Air pollution,

influenza, and mortality in New York City, Arch. Environ. Health 15:430-438, 1967.

47. Greenburg, L., Field, F., Reed, J. I., and Erhardt, C. L.: Air pollution and morbidity in New York City, J.A.M.A. 182:161-164, 1962.

48. Greenburg, L., Jacobs, M. B., Drolette, B. M., Field, F., and Braverman, M. M.: Report of an air pollution incident in New York City, November, 1953, Public Health Rep. 77:7-16, 1962.

49. Griswold, S. S., Chambers, L. A., and Motley, H. L.: Report of a case of exposure to high ozone concentrations for two hours, Arch. Industr. Health 15:108-110, 1957.

50. Haagen-Smit, A. J.: Carbon monoxide levels in city driving, Arch. Environ. Health 12: 548-551, 1966.

51. Haagen-Smit, A. J., and Fox, M. M.: Ozone formation in photochemical oxidation of organic substances, Industr. Engin. Chem. 48: 1484-1486, 1956.

52. Haagen-Smit, A. J., and Wayne, L. G.: Atmospheric reactions and scavenging processes, in Air pollution, vol. 1, New York, 1968, Academic Press, Inc., chap. 6, pp. 149-186.

53. Haggard, H. W.: Studies in carbon monoxide asphyxia. I. The behavior of the heart, Amer. J. Physiol. 56:390-403, 1921.

54. Haldane, J.: The action of carbonic oxide on man, J. Physiol. 18:430-462, 1895.

55. Hatch, T., and Hemeon, W. C. L.: Influence of particle size in dust exposure, J. Industr. Hyg. Toxic. 30:172-180, 1948.

56. Hickey, R. J., Schoff, E. P., and Clelland, R. C.: Relationship between air pollution and certain chronic disease death rates, Arch. Environ. Health 15:728-738, 1967.

57. Holland, W. W., and Reid, D. D.: The urban factor in chronic bronchitis, Lancet 1: 445-448, 1965.

58. Holland, W. W., Reid, D. D., Seltser, R., and Stone, R. W.: Respiratory disease in England and the United States, Arch. Environ. Health 10:338-343, 1965.

59. Ipsen, J., Deane, M., and Ingenito, R. E.: Relationships of acute respiratory disease to atmospheric pollution and meteorological conditions, Arch. Environ. Health 18:462-472, 1969.

60. Ishikawa, S., Bowden, D. H., Fisher, V.,

and Wyatt, J. P.: The "'emphysema profile" in two midwestern cities in North America, Arch. Environ. Health **18**:660-666, 1969.

61. Jaffe, L. S.: The biological effects of photochemical air pollutants on man and animals, Amer. J. Public Health **57**:1269-1277, 1967.

62. Jones, R. H., Ellicott, M. F., Cadigan, J. B., and Gaensler, E. A.: The relationship between alveolar and blood carbon monoxide concentrations during breathholding, J. Lab. Clin. Med. **51**:553-564, 1958.

63. Kjeldsen, K. Thesis: Smoking and atherosclerosis: Investigations on the significance of the carbon monoxide content in tobacco smoke in atherogenesis, Univ. of Copenhagen, 1969.

64. Landahl, H. D., and Herrmann, R. G.: On the retention of air-borne particulates in the human lung. J. Industr. Hygiene Toxic **30**: 181-188, 1948.

65. Lawther, P. J.: Effects of inhalation of sulphur dioxide on respiration and pulse-rate in normal subjects, Lancet **2**:745-748, 1955.

66. Lilienthal, J. L., and Fugitt, C. H.: The effect of low concentrations of carboxyhemoglobin on the "altitude tolerance" of man, Amer. J. Physiol. **145**:359-364, 1946.

67. Lilienthal, J. L., and Pine, M. B.: The effect of oxygen pressure on the uptake of carbon monocide by man at sea level and at altitude, Amer. J. Physiol. **145**:346-350, 1946.

68. Loudon, R. G., and Kilpatrick, J. F.: Air pollution, weather, and cough, Arch. Environ. Health **18**:641-645, 1969.

69. McCarroll, J., and Bradley, W.: Excess mortality as an indicator of health effects of air pollution, Amer. J. Public Health **56**:1933-1942, 1966.

70. McCarroll, J., Cassell, E. J., Wolter, D. W., Mountain, J. D., and Diamond, J. R.: Air pollution and illness in a normal urban population, Arch. Environ. Health **14**:178-184, 1967.

71. McFarland, R. A., Roughton, F. J. W., Halperin, M. H., and Niven, J. I.: The effects of carbon monoxide and altitude on visual thresholds, J. Aviat. Med. **15**:381-394, 1944.

72. Miranda, J. M., Konopinski, V. J., and Larsen, R. I.: Carbon monoxide control in a high highway tunnel, Arch. Environ. Health **15**:16-25, 1967.

73. Motley, H. L., Smart, R. H., and Leftwich, C. I.: Effect of polluted Los Angeles air (smog) on lung volume measurements, J.A.M.A. 171:1469-1477, 1959.

74. National Air Quality Criteria Advisory Committee. Air Quality Criteria for Particulate Matter. United States Department of Health, Education, and Welfare. Public Health Service, January, 1969.

75. National Air Quality Criteria Advisory Committee. Air Quality Criteria for Sulfur Oxides. United States Department of Health, Education, and Welfare. Public Health Service, January, 1969.

76. Neiburger, M.: Meterological aspects of air pollution, Arch. Environ. Health 14:41-45, 1967.

77. Oshima, Y., Ishizaki, T., Miyamoto, T., Shimizu, T., Shida, T., and Kabe, J.: Air pollution and respiratory disease in the Tokyo-Yokohama area, Amer. Rev. Resp. Dis. 90:572-581, 1964.

78. Pace, N., Consolazia, W. V., White, W. A., Jr., and Behnke, A.: Formulation of the principal factors affecting the rate of uptake of carbon monoxide by man, Amer. J. Physiol. 147:352-359, 1946.

79. Phelps, H. W.: Follow-up studies in Tokyo-Yokohama respiratory disease, Arch. Environ. Health 10:143-147, 1965.

80. Prindle, R. A., Wright, G. W., McCaldin, R. O., Marcus, S. C., Lloyd, T. C., and Bye, W. E.: Comparison of pulmonary function and other parameters in two communities with widely different air pollution levels, Amer. J. Public Health 53:200-217, 1963.

81. Ramsey, J. M.: Concentrations of carbon monoxide at traffic intersections in Dayton, Ohio. Arch. Environ. Health 13:44-46, 1966.

82. Reinke, W. A.: Multivariable and dynamic air pollution models, Arch. Environ. Health 18:481-484, 1969.

83. Rouché, B.: Eleven blue men, Boston, 1953, Little, Brown & Company.

84. Sawicki, E.: Airborne carcinogens and allied compounds, Arch. Environ. Health 14:46-53, 1967.

85. Schulte, J. H.: Effects of mild carbon monoxide intoxication, Arch. Environ. Health 7:524-530, 1963.

86. Shafer, N., Smilay, M. G., and MacMillan, F. P.: Primary myocardial disease in man re-

sulting from acute carbon monoxide poisoning, Amer. J. Med. **38**:316-320, 1965.

87. Snell, R. E., and Luchsinger, P. C.: Effects of sulfur dioxide on expiratory flow rates and total respiratory resistance in normal human subjects, Arch. Environ. Health **18**:693-698, 1969.

88. Speizer, F. E., and Frank, N. R.: A comparison of changes in pulmonary flow resistance in healthy volunteers acutely exposed to SO_2 by mouth and by nose, Brit. J. Industr. Med. **23**:75-79, 1966.

89. Spodnik, M. J., Jr., Cushman, G. D., Kerr, D. H., Blide, R. W., and Spicer, W. S., Jr.: Effects of environment on respiratory function, Arch. Environ. Health **13**:243-254, 1966.

90. Sterling, T. D., Pollack, S. V., and Weinkam, J.: Measuring the effect of air pollution on urban morbidity, Arch. Environ. Health **18**:485-494, 1969.

91. Stern, A. C., editor: Air pollution vol. II, Analysis, monitoring, and survey, New York, 1968, Academic Press, Inc.

92. Strandberg, L. G.: SO_2 absorption in the respiratory tract, Arch. Environ. Health **9**: 160-166, 1964.

93. Surgeon General's Advisory Committee on Smoking and Health. Smoking and Health: Report of the advisory committee to the Surgeon General of the Public Health Service. United States Department of Health, Education, and Welfare. Public Health Service, No. 1103.

94. Thurlbeck, W. M.: The geographic pathology of pulmonary emphysema and chronic bronchitis, Arch. Environ. Health **14**:16-20, 1967.

95. Toyama, T.: Air pollution and its health effects in Japan, Arch. Environ. Health **8**: 153-173, 1964.

96. Trouton, D., and Eysenck, H. J.: The effects of drugs on behavior, Eysenck, H. J., editor: Handbook of abnormal physiology, New York, 1961, Basic Books, Inc. pp. 634-696.

97. Ury, H. K., and Hexter, A. C.: Relating photochemical pollution to human physiological reactions under controlled conditions, Arch. Environ. Health **18**:473-480, 1969.

98. Verma, M. P., Schilling, F. J., and Becker, W. H.: Epidemiological study of illness absences in relation to air pollution, Arch. Environ. Health **18**:536-543, 1969.

99. Waller, R. E., and Lawther, P. J.: Further observations on London fog, Brit. Med. J. **2:**1473-1475, 1957.

100. Wanstrup, J., Kjeldsen, K., and Astrup, P.: Acceleration of spontaneous atherosclerotic changes in rabbits by a moderate, prolonged carbon monoxide exposure, Acta Path. Microbiol. Scand. **75:**353-362, 1969.

101. Weill, H., Ziskind, M. M., Derbes, V., Lewis, R., Horton, R. J. M., and McCaldin, R. O.: Further observations on New Orleans asthma, Arch. Environ. Health **8:**184-187, 1964.

102. Wilson, I. B., and LaMer, V. K.: The retention of aerosol particles in the human respiratory tract as a function of particle radius, J. Industr. Hyg. Toxic. **30:**265-280, 1948.

103. Winkelstein, W., Jr., and Kantor, S.: Stomach cancer. Positive associate with particulate air pollution, Arch. Environ. Health **18:** 544-547, 1969.

104. Winkelstein, W., Jr., Kantor, S., Davis, E. W., Maneri, C. S., and Mosher, W. E.: The relationship of air pollution and economic status to total mortality and selected respiratory system mortality in men. I. Suspended particulates, Arch. Environ. Health **14:**162-171, 1967.

105. Winkelstein, W., Jr., Kantor, S., Davis, E. W., Maneri, C. S., and Mosher, W. E.: The relationship of air pollution and economic status to total mortality and selected respiratory system mortality in men. II. Oxides of sulfur, Arch. Environ. Health **16:**401-405, 1968.

106. Wright, G. Air pollution: A conference report, Public Health Rep. **75:**1179-1180, 1960.

107. Xinteras, C., Johnson, B. L., Ulrich, C. E., Terrill, R. E., and Sobecki, M. F.: Application of the evoked response technique in air pollution toxicology, Toxic. Appl. Pharmacol. **8:**77-87, 1966.

108. Young, W. A., Shaw, D. B., and Bates, D. V.: Effects of low concentrations of ozone on pulmonary function, J. Appl. Physiol. **19:** 765-768, 1964.

The Correlation of Polluted Air with Tree Growth and Lung Disease in Humans

JAMES A. HEIMBACH, JR.

Department of Meteorology, The University of Oklahoma,
Norman, Oklahoma, U.S.A.

Abstract—In Altoona, Pennsylvania, a study was done to correlate the incidence of lung cancer and pulmonary emphysema to air pollution. Spatial distributions of relative sulfur dioxide concentrations were calculated for Altoona through a pollution model based on the Pasquill–Gifford equation of diffusion using the city's climatological and source distributions. A statistical analysis was done comparing the incidence of lung disease with concentrations of sulfur dioxide. A positive correlation was indicated. Lateral growth rates of trees throughout the city were analyzed to compare with the concentration patterns obtained from the pollution model. Areas showing past stunting of tree growth coincided reasonably with those areas shown numerically to be polluted.

1. INTRODUCTION

THE NASHVILLE air pollution study,[1] run from 1957 to 1959, showed a relation between air pollution and bronchial asthma, anthracosis, mortality and morbidity. A study done by GILBERT[2] correlated the lack of certain types of lichens to air pollution in the Tyne Valley in England. THOMAS[3] listed some of the effects of specific types of air pollution to plant as well as animal health. To further investigate some possible harm that air pollution could have on Man, a correlation between lung disease and air pollution was sought. This endeavor was broken into three parts: investigating all cases of typical lung diseases based on hospital records over a period of one year, comparing these patterns with spatial distributions of relative sulfur dioxide concentrations predicted by a pollution model, and checking the results of the modelling by analyzing tree growth throughout a city.

It was thought best to use a small city which had at one time a severe pollution problem. The city was to be small enough to make a complete survey of lung diseases possible, as well as a complete tree growth analysis, but still large enough to give a meaningful trend. A changing pollution profile was sought to provide differing growth rates in trees, as will be explained later. Altoona, Pennsylvania, was chosen.

2. THE CITY

Altoona is about 100 miles to the east of Pittsburgh in the northern part of the Appalachian Mountains. The city was established about 1850 as a western maintenance center for the Pennsylvania Railroad (PRR), and grew into the main maintenance depot where all heavy repairs were done as well as the complete production of steam locomotives.

On either side of Logan Valley, where Altoona is located, are hills up to 1200 ft higher than the valley floor.[4] These southwest to northeast oriented hills would tend to influence the winds to prevail from the southwest, as verified when analyzing old railroad weather data. Because of the nature of the industry, there was heavy air pollution in Altoona, the peak being around 1945. Between 1950–1960, however, air pollution from coal burning has decreased to almost none because of the diesel engine replacing the steam locomotive and the fact that the railroad industry in Altoona had been drastically cut back.

3. DATA COLLECTION

Pollution sources

All the sources considered belonged to the Pennsylvania Railroad. There were three categories of sources: power generating stacks; line sources, i.e. railroad tracks; and area sources made up of switching yards and maintenance shops. During the era of the steam engine, coal burning accounted for practically all of the combustion products. Interviews with railroad men disclosed that anthracite coal was used and the sulfur content averaged from two to four per cent.

There were seven power stacks in Altoona which were considered (see Fig. 2). A main line of the PRR went through Altoona from Harrisburg to Pittsburgh with a small branch going to Hollidaysburg. Shops and switching yards were located in the northeast and central portions of the city.

Many records of the amount of coal burned by the various sources were destroyed by summer, 1969, when this investigation was begun. Estimates of the amounts burned by those sources for which there were no records available, were obtained from the memories of the older railroad workers. Interviews with railroad veterans also gave estimates of the plume rise. Plume rise is the lifting of the effluent above the source due to buoyancy and upward velocity. This added height varied from 15 per cent of the source height for power stacks to 50 per cent for the steam locomotives for normal weather conditions.

Tree data

Tree cores were taken throughout Altoona. An attempt was made to sample only maple trees because it is believed by some[5] that maples are more sensitive to sulfur dioxide pollution. In many areas, however, there were no maples to be found. In some areas, where there was previously very heavy pollution, the only older trees available were the *Ailanthus*, or there were no trees older than 1955 to be found.

Disease data

The types of lung disease used in the study were lung cancer and pulmonary emphysema. These were investigated because they can be a result of continued irritation.[6] The two public hospitals, the Altoona and Mercy Hospitals, had their records checked for all lung cancer and emphysema cases which were entered during the year 1965. This year was settled upon for several reasons. First, better records are available for recent years because of increased management efficiency at both hospitals. Second, medical science has developed more accurate diagnoses of lung diseases, and third, the smoking habits of the patients must be considered. Prior to 1960, little concern over smoking was evident, and even in 1965, only 70 per cent of the cases considered had anything entered on their hospital records about smoking.

The residence where each patient was living at the time of diagnosis was recorded and later converted to a grid address. This address was assumed to be where the patient lived from at least 1945, a safe assumption because the older Altoonians are not very nomadic. Only four of 171 emphysema and lung cancer cases were under 45—with ages 6, 26, 40 and 44—having emphysema. Also recorded were sex, extent of ailment, age of patient, smoking habits, source of data, income, place of employment, season of admission, and population density where the patient lived. Only those cases which resided within the city limits were considered.

4. COMPUTER ANALYSIS

The analysis done by computers was in three parts: sorting of disease and tree growth data, numerical air pollution modelling, and application of the results of this modelling to disease data for correlation purposes. Most of the computation was done on the CDC 6600 located at NCAR, Boulder, Colorado. The data organization was an elementary exercise of classifying and plotting grid locations of incidence in the case of disease data as shown in Figs. 5–7. With the tree data, this involved finding comparative growth rates and plotting these on a printed representation of the city (see Fig. 4). For this portion, as well as for the rest of the study, the city was sectioned into a 53 by 108 grid.

As there were no air quality control data for Altoona, a theoretical method of estimating air pollution values had to be used. The Pasquill–Gifford diffusion model, as explained in the next section, was applied to individual sources within Altoona. Each area of the 53 by 108 grid had the contributions from each source of three source and weather configurations summed. The model was run for an up valley wind using all sources mentioned in the previous section. Also, average relative pollution concentrations for two fields were estimated using a yearly sixteen point wind climatology with all sources and with only the seven main power stacks. In all three pollution fields the point source Pasquill–Gifford model was applied to a multi-source situation to obtain a dense grid of values.

The fields of pollution computed using the yearly climatology were applied to correlate the incidence of emphysema and lung cancer to relative values of air pollution. Because pollution values and disease locations were referenced to the same grid, statistical analyses simply involved searching for the respective pollution value of the address of each disease case. Later in the study, when a map giving population densities of Altoona was obtained, an analysis was done to check for randomness of disease incidence with respect to population.

5. POLLUTION MODELLING

The working equation used for computing pollution values is a special form of the Pasquill–Gifford diffusion treatment where pollution concentrations are found at the surface.[7]

$$\chi[x,y,0;H] = \frac{Q}{\pi\sigma_y\sigma_z u} \exp\left[-y^2/2\sigma_y{}^2 - H^2/2\sigma_z{}^2\right]. \tag{1}$$

In this approach, air pollution concentration is assumed to be bi-normally distributed with respect to vertical, z, and across wind, y, directions. The density of a normal distribution as a function of y only is denoted by $f(y)$

$$f(y) = \frac{1}{\sqrt{2\pi}\,\sigma_y} \exp\left[-y^2/2\sigma_y{}^2\right].$$

The density as a function of z only is $g(z)$

$$g(z) = \frac{1}{\sqrt{2\pi}\,\sigma_z} \exp\left[-z^2/2\sigma_z{}^2\right].$$

E

162

The standard deviations of these distributions are σ_y, σ_z. The point source is assumed to be continuously emitting, i.e. steady state conditions are assumed, making stretching in the downwind direction a function of the inverse of average wind velocity, $1/u$. Multiplying these three contributions of diffusion together, and allowing for the plume to be reflected at the earth's surface from a source at an effective height H, gives:

$$\text{Joint density at surface} = \frac{1}{\pi\sigma_y\sigma_z u} \ \exp\left[-y^2/2\sigma_y^2 - H^2/2\sigma_z^2\right].$$

The concentration of pollution, χ, at a point $x, y, z = 0$; due to a source with effective height H, and emission rate Q, is $\chi[x,y,0;H]$. This is Q times joint density, giving equation (1).

The σ's are a function of downwind distance, x. As the plume travels downwind it widens vertically and horizontally having the σ's increase as x increases. During unstable atmospheric conditions, the increased turbulence would have the σ's increasing at a faster rate than for stable conditions. Graphed values of standard deviation as a function of downwind distance have been found empirically for six stability conditions.[7,8] These graphs were approximated by exponential functions which were used to give values of σ_y and σ_z for any x when the model was applied.

Although equation (1) can be applied to line and area sources with some reworking, it was more economical, in terms of computer time, to approximate the line and area sources by a series of point sources. Due to the roughness of the source data, and the uncertainty of mapping long-term averages of pollution, relative densities of pollution were mapped. These had the same order of magnitude as one could expect of actual long term averages.

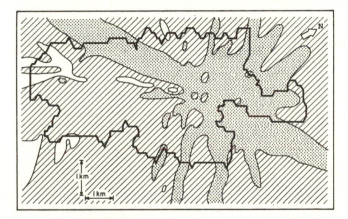

Fig. 1. A yearly average pollution field computed using the seven main power stacks of the Pennsylvania Railroad with the yearly wind climatology. The dotted areas represent values greater than 10^{-4} g/m³, the hatched areas, greater than 10^{-5} g/m³, and the other areas represent values of less than 10^{-5} g/m³. This shading scheme is used for the other pollution figures. On this as well as all other figures, the map is stretched slightly along the northwest to southeast direction. This was done because output from line printing required some stretching in this direction to preserve resolution.

Three results of the modelling are presented. The first, Fig. 1, used only the seven power stacks as sources with winds and their directions averaged from five years of weather records. Each wind direction contributed to the yearly average pollution value with a weight equal to the amount of time the wind blew from that direction. The second, Fig. 2, used all pollution sources with the same climatology, and the third, Fig. 3, used all the sources with the prevailing up-valley wind only. The wind climatology has sixteen compass points and for each direction an average atmospheric stability was hypothesized. For example, winds up the valley (from the southwest) were associated with neutral stabilities. There are some similarities among the three results worth pointing out. All show heavy pollution values to the northeastern portions of the figures. These areas are known locally as Juniata, East Altoona, and Greenwood. Also, all figures show a heavy pollution ribbon from the center of the city to the west-south-west. The narrow bands of Figs. 1 and 2, which radiate from the center, are attributable to the sixteen discrete directions of the wind climatology.

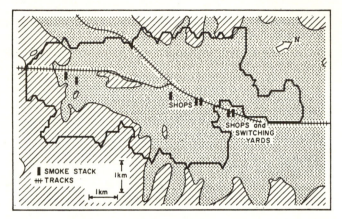

FIG. 2. The yearly average pollution field computed using all railroad sources with the yearly wind climatology. Shown also in this figure are the positions of the stacks, area sources, and line sources used in the study.

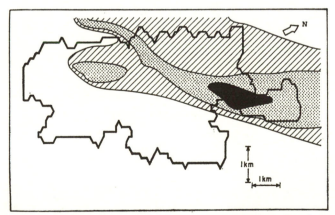

FIG. 3. The pollution field obtained using all railroad sources with prevailing upvalley wind. The black area represents relative values greater than 10^{-3} g/m³.

164

6. TREE ANALYSIS

There are so many variables to consider when synthesizing a pollution field by a computer that some type of alternate method should be used to obtain a pollution profile. A combination of the two methods would make possible a better idea of the actual situation. As pollution measurements by instrument were nonexistent in Altoona, some type of indicator had to be used which was present in the city since 1945. The indicator used was the growth rate of trees.

It is believed by some [9] that air pollution stunts tree growth. If this is so, an area which had heavy air pollution in the past would have trees growing faster in the years of suppressed air pollution. There are, of course, other factors involved with tree growth such as rainfall, temperature, disease, damage by children, crowding by other trees, paving close to the tree, and the age of the tree. By measuring the amount of growth over several years and comparing this growth to an equal span of time, the short term effects of rainfall could be filtered out. If the other factors mentioned above, excepting age, were assumed constant and if no exceptionally old trees were analyzed, then the differing growth rates could be attributed to differing pollution. What was looked for was a rate of change of size with respect to time:

$$\frac{\delta \text{(size)}}{\delta t}.$$

A gradient of this change of size with respect to time, or $\frac{\delta}{\delta x} \frac{\delta \text{(size)}}{\delta t}$ would be meaningless because there were several types of trees considered under many growing conditions. To get around this, each tree was used as its own control and the following ratio called "FACTOR" was defined.

$$\text{FACTOR} = \frac{\text{Growth of tree in early period}}{\text{Growth of tree in later period}}$$

$$= \left[\frac{\delta \text{(size)}}{\delta t}\right]_{\text{period } A} \bigg/ \left[\frac{\delta \text{(size)}}{\delta t}\right]_{\text{period } B}. \tag{2}$$

It is the gradient of this "factor" which provides an idea of the past pollution profile. If "factor" gives a result equal to 1, then, either there would be no pollution during the two periods, or the pollution rate has remained constant. If the result is less than 1, then there would be more pollution during the early period than the later. A "factor" greater than 1 would imply more pollution during the later period or something else which would cause the tree to grow slower in the later period. Significantly less rain for several years, damage by children, or insect damage during the later period are some examples of this.

Altoona had heavy air pollution before 1950. Between 1950–1960, the pollution decreased to a fraction of what it was, remaining at this low level through 1969. The "factors", as defined in equation (2), should be less than unity over areas of the city which were polluted, particularly when comparing the growth of 1945–1955 to later years.

Figure 4 shows the contours for the factors derived by comparing 1945–1955's growth to 1955–1965's growth. Areas of large stunting occurred to the northeast of the city and along the Altoona to Pittsburgh railroad tracks. A patch of heavy stunting is to the southwest of the city, probably attributable to the two power stacks in that area. Areas of light and heavy stunting correspond closely to those areas shown by the theoretical method to be polluted.

165

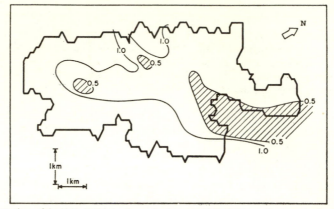

Fig. 4. A comparison of the 1945–1955 lateral growth of trees to 1955–1965 growth. Areas having no trees older than 1945 were included in the very stunted areas (hatched). Areas having stunted tree growth prior to 1955 correspond with areas shown numerically to be polluted.

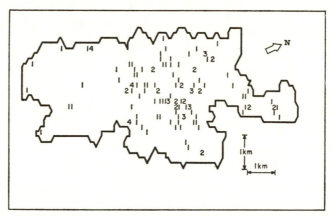

Fig. 5. The locations of all emphysema cases living within Altoona during 1965. All cases were diagnosed in the two local hospitals. Each integer represents the number of cases residing within each grid area.

7. DISEASE ANALYSIS

All locations of cases of lung cancer and emphysema were plotted on an outline of Altoona. (See Fig. 5 for emphysema). Both plots showed a grouping of cases near the center of the city and a lack of cases to the southwest. Some of this grouping was due to a difference of population density, but, as shown later, there was some other factor involved in this grouping. The definitely nonsmoking lung cancer (Fig. 6) and emphysema (Fig. 7) are also plotted. The plot of definitely nonsmoking lung cancer cases shows an interesting grouping around the Altoona to Pittsburgh railroad tracks. There was only one case not in this area and it turned out that this was the only nonsmoking lung cancer patient who did not work for the railroad.

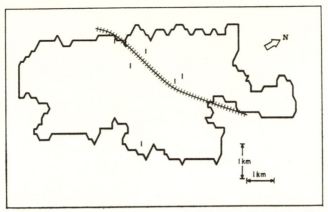

FIG. 6. The locations of definitely nonsmoking lung cancer patients reported in 1965. A section of the Altoona to Pittsburgh railroad track was drawn to show the grouping of five out of the six cases around these tracks. The case to the southeast, not in this grouping, was the only nonsmoking lung cancer patient who was not a railroad employee.

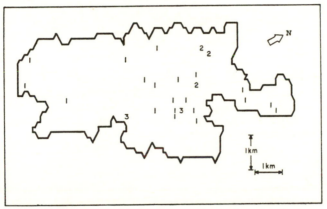

FIG. 7. The locations of definitely nonsmoking emphysema patients.

Comparing average pollution values

The mean pollution values of grid areas having an incidence of a type of lung disease were compared to the mean values of those points not having any lung disease reported. A version of Student's *t*-test was used to test for significance of difference.[10] This test was used because the parameter variances of the distributions of pollution values could not be assumed to be equal. The parameter is defined as follows:

$$t_s = \frac{[\bar{y}_1 - \bar{y}_2] - [\mu_1 - \mu_2]}{[s_1{}^2/n_1 + s_2{}^2/n_2]^{\frac{1}{2}}}. \tag{3}$$

The $\bar{y}$'s are the sample means and the null hypothesis has the parameter means, μ's, equal. The s^2's are the sample variances and the n's are the number of samples. The two pollution fields computed using the yearly climatology, were used to obtain the pollution values. All cases, as well as only nonsmoking cases of the two lung diseases, had their means compared with the means of those points not having any cases reported. No grid areas outside the city's boundaries were considered. The parameter is compared with a weighted value of t'_a to determine the significance of difference:[10]

$$t'_a = \frac{t_{a[v_1]}s_1{}^2/n_1 + t_{a[v_2]}s_2{}^2/n_2}{s_1{}^2/n_1 + s_2{}^2/n_2}.$$

The subscript a is the critical probability, which is used to define a significant difference, and the $t_{a[v]}$'s are found in statistical tables. The degrees of freedom are the v's.

It turned out that all the means of the sets of points showing an incidence of lung disease were greater than those not having any incidence, whether considering all cases or nonsmoking cases only. Table 1 summarizes the results of the testing of these means.

TABLE 1

SUMMARIZED RESULTS OF STUDENT'S T-TEST FOR SIGNIFICANCE
OF DIFFERENCE BETWEEN MEAN POLLUTION CONCENTRATIONS

	Pollution field computed using:	
	all sources	seven main sources
All emphysema	0·001	0·001
All lung cancer	0·05	0·4
Nonsmoking emphysema	0·05	0·1
Nonsmoking lung cancer	0·3	0·2

Figure 8 shows a plot of the ratios of the number of diseased grid areas over the non-diseased areas for classed pollution values. The pollution field of Fig. 2 was used for this plot. Lung cancer, and to a greater degree, emphysema, show an increase in this ratio toward higher pollution values. This illustrates that areas with higher pollution values had a greater chance of having a residing lung patient.

Analysis of variance

As an alternate method to test for significance of difference among the mean pollution values of diseased areas vs. nondiseased areas, an analysis of variance (Anova) was done.[11] The data were stratified into three groups: areas having no incidence of lung disease, areas having lung cancer, and areas having emphysema. Four anovas were done on the pollution values of these three groups. The pollution values were obtained from the pollution fields of Figs. 1 and 2. Two anovas used only the definite nonsmokers and two used all cases reported. Each of the anovas showed a significant variance among the mean pollution values of the groups. The significance levels obtained were 0·05 for both nonsmoker anovas, and 0·001 for both anovas which used all cases reported.

Population density

There was no correction for varying population densities in the above analyses. To test if the disease cases were randomly distributed with respect to population density, the city was divided into 65 areas each containing 1000 people. The areas were based on the 1960 census. The numbers of cases of lung cancer and emphysema occurring in each area were

168

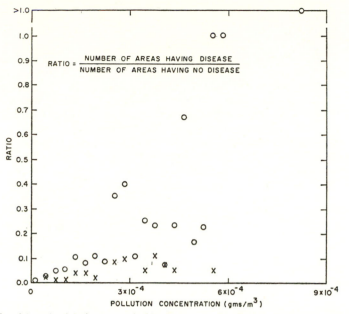

Fig. 8. A plot of the ratio of the frequency of grid areas having an incidence of lung cancer and emphysema to the frequency of nondiseased areas for classed pollution values. The pollution field of Fig. 2 was used for this figure. The circles indicate ratios for emphysema.

plotted vs. frequency of each number. These distributions were compared with the expected Poisson frequencies and both showed an excess of frequencies on the tails, demonstrating "clumping".[10] That is, there were more areas containing small and large numbers of cases than expected, whereas the frequencies in the middle of the distributions were less than expected. Clumping can be interpreted as meaning there was some influence in Altoona encouraging the incidence of these diseases. Whether this was air pollution or not cannot be decided solely by this analysis.

A chi-squared goodness of fit test was applied to determine the significance of difference between the actual frequencies and the expected frequencies of disease occurrence within these areas of equal population. The distribution of lung cancer was found to be different at a significance level of 0·2, and the emphysema at a level of 0·005. For emphysema, this can be interpreted as meaning that there was less than one chance in 200 that the emphysema cases could be distributed as they were, in a random fashion. This implies that there was some ordering imposed on the disease incidence.

8. SUMMARY AND CONCLUSIONS

Although the time of heavy pollution and time of disease occurrence were ten years apart, this project has found evidence that there was some residual effect of the past air pollution of Altoona on certain lung ailments. By fabricating several numerical pollution fields and checking their combined validity with tree growth analysis, long term average values of air pollution were obtained. Values of air pollution taken from two of these fields

169

were used to see if grid points having an incidence of lung cancer or pulmonary emphysema had higher means than those not having any incidence. In all cases the diseased areas had higher pollution averages.

The effects of population density on the grouping of disease cases was shown to be offset by some factor, presumably air pollution. Analysis showed that areas tended to have either no cases of emphysema or lung cancer or many cases, i.e. a "clumping" of cases in certain areas. Through use of the t-test of means and analysis of variance, it was found that those areas where the disease cases clumped tended to be areas shown numerically to be polluted.

For the number of lung ailments obtained in 1965 (a total of 171 cases), the results have been most encouraging. Further work is planned along this line using a larger city. The pollution model will be updated to include topography, work schedules of factories, and the urban heat island effect. Future plans also include short term effects of several types of air pollution on lung ailments.

Acknowledgements—I would like to thank Y. SASAKI, G. A. EDDY and E. L. RICE for their help and encouragement while working on this project. Acknowledgment is also made to the National Center for Atmospheric Research, which is sponsored by the National Science Foundation, for use of its Control Data 6600 computer.

REFERENCES

1. J. D. WILLIAMS and N. G. EDMISTEN, *An Air Resource Management Plan for the Nashville Metropolitan Area*. U.S. Department of Health, Education, and Welfare, Cincinnati, Ohio, 1965, pp. 86–90.
2. O. C. GILBERT, Lichens as indicators of air pollution in the Tyne Valley, *Ecology and the Industrial Society*, A Symposium of the British Ecological Society (Symposium No. 15, 13–16 April 1964).
3. M. D. THOMAS, Effects of air pollution, *Ecology and the Industrial Society*, A Symposium of the British Ecological Society (Symposium No. 15, 13–16 April, 1964).
4. Blair County Planning Commission, Historical background and physiography, Report No. 12 of the Blair County Planning Commission, Hollidaysburg, Pennsylvania, 1967, pp. 4–37.
5. W. E. HERMAN, From a personal interview, Department of Forests and Waters, Altoona, Pennsylvania, 1969.
6. S. W. TROMP, *Medical Biometeorology*, pp. 479–480 and 499–501. Elsevier, New York, 1963.
7. United States Atomic Energy Commission, *Meteorology and Atomic Energy*, p. 99. U.S. Department of Commerce, Springfield, Virginia, 1969.
8. D. B. TURNER, *Workbook of Atmospheric Dispersion Estimates*, pp. 8–9. U.S. Department of Health, Education and Welfare, Cincinnati, 1969.
9. L. J. BATTAN, *The Unclean Sky*, pp. 74–77. Doubleday, Garden City, New York, 1966.
10. R. R. SOKAL and F. J. ROHLF, *Biometry*, pp. 90–94, 221, 374. W. H. Freeman, San Francisco, 1969.
11. P. G. HOEL, *Introduction to Mathematical Statistics*, pp. 299–325. John Wiley, New York, 1962.

Tests to Assess Effects of Low Levels
of Air Pollutants on Human Health

Benjamin G. Ferris, Jr., MD

A VARIETY of tests and techniques have
been used to assess the effects of air pollutants on human health. Some of these have
helped to define the effects of relatively high
levels of pollution. In our attempts to identify the effects of low levels of pollutants,
some of the tests have not been demonstrated to have sufficient sensitivity, and others
have been shown to be relatively ineffective.
It does, therefore, seem appropriate to review the various tests and techniques and
to indicate the more productive existing
methods and where more studies are needed
to determine the sensitivity and specificity
of the methods. Whatever is said below implies that we know and have good measurements of the level of air pollution so that
results of tests can be related to levels of
pollution.

Mortality

Studies of mortality have been useful in
the past to identify areas of severe hazard or
to help identify toxic pollutants. Such
studies have been helpful in demonstrating
effects during periods in which air pollutants have risen to relatively high levels for
brief periods of time.[1,2] On the other hand,

when effects of chronic low levels of air pollutants are sought, mortality data are much less effective indicators. This results from their lower sensitivity because exposures are varied, and there may be a number of complicating factors such as cigarette smoking or occupational exposures. The end point occurs a long time after exposure started, and although clearly defined, it is measuring the cumulative results of a great many years; and migration may have occurred. This is not the place to discuss in detail all of the problems associated with such methods. As mortality is presently examined, it is too crude for studies on the chronic effects of low levels of air pollution.

When large population groups can be studied so that the number of deaths per day becomes large, more sophisticated statistical analyses can be used, such as variance or stepwise regression analysis. The deaths on a given day can be related not only to levels of pollution but also to climatic factors such as temperature, humidity (relative and absolute), precipitation, wind, etc. Deviations from various moving averages can also be used. Lag times can be introduced. Rate of change of climatic condition can be examined also, as well as absolute value. Studies of this sort are indicated and should be extended beyond the effects of air pollutants as some have done.[3]

Morbidity and Impairment of Physiological Function.—Studies of morbidity intuitively would seem to be more useful than studies of mortality in assessing the effects of chronic exposure to low levels of pollutants. This, in fact, has been the case and is particularly evident when alterations in function can be determined. Along with morbidity, we include specific alterations in function that may not necessarily be associated with subjective complaints, as, for example, a lowered vital capacity (FVC) or (one second) forced expiratory volume ($FEV_{1.0}$).

Populations selected may be communities or occupational groups with special types of exposures. In studies of this type, one must select a suitable population for study. If a population is so large that it is impractical to study the entire population, random sampling should be done to obtain a representative group. Weighted samples or stratified samples may at times be selected. Whichever techniques are used will be indicated by the questions to be asked, the objectives of the survey, and the number of subjects needed in the various categories to demonstrate a significant difference.

Subjective Methods.—Diaries have been used by some investigators.[4-7] In these, the subjects have recorded their symptoms daily or interviewers have made regular visits. The symptoms elicited in turn have been related to the levels of pollution. In general, these have tended to have limited value in part because some subjects respond to one material and others to another or the group may respond together but react to different pollutants on different days.[5] Errors of omission or addition also occur, but, hopefully, these cancel out. This still requires retrospection and remembering to keep the diary daily or recalling the symptoms of the previous week. Unless the method of recording can be improved upon or a particularly susceptible group could be identified for follow-up, this technique appears to be of limited use.

In addition, there should be more complete documentation of exposures either by means of personal dosimeters or more monitoring stations. Some of the variation in responses may be related to changes in wind directions so that one group is exposed and the other is not. This is particularly important when there is a point source.

Questionnaires have had extensive use in a number of countries. The British developed and have been the advocators of a standardized form. This is certainly the first step if comparisons are to be made between studies. If such comparisons are not the purpose of the study, then the questionnaire can be adapted to that particular study. Observer differences may be a real problem, but when attention has been paid to using the same questionnaire in a standardized manner, observer differences are minimal and are what one might expect by chance variation alone. A greater problem arises when the questionnaire is used in different cultures and translated into another language. Words take on different meanings, and the culture itself may condition the response. Thus, differences seen between studies must consider the possible effect of culture; and cultural differences may well vitiate the comparison. This is an area where studies are needed to assess the effect of culture on responses to questionnaires. It may be that reliance will have to be put on more objective tests when international comparisons are to be made.

Results from the questionnaires have been particularly useful in areas where there have been relatively high levels of air pollution as in Great Britain. The questionnaire also permits obtaining much necessary basic information that is used in comparing or standardizing studies such as ethnic or cultural background, socioeconomic level, level of formal education, places of residence, occupational histories, and tobacco smoking habits, with particular emphasis on cigarette smoking; age and height are also obtained. These are less influenced by cultural patterns, but ethnic factors may be important in using prediction formulae where sitting height standing-height ratios may differ markedly between ethnic groups.[8]

The usefulness of the questionnaire has seemed to be limited when used to assess the effects of air pollution at the levels encountered in North America. What is needed are more studies of a standardized nature in more areas of North America and with more of the different levels of air pollution represented. Much of the poor sensitivity of the questionnaire may result from the competing effect of cigarette smoking. This "background noise" may be so great that the small signal received from the relatively low levels of air pollution cannot be clearly identified, or much larger population groups must be studied. This latter will be extremely costly.

Studies of nonsmoking populations, specifically children, have been done because of the competing effect of cigarette smoking. Absences from school of first and second graders have been followed without demonstrating significant differences between areas with different levels of pollution in United States and Canada.[9,10] Prolonged absences (more than one week), however, have been reported to be more frequent in areas of high pollution in Great Britain.[11] Symptom prevalences as compared with levels of air pollution have also been followed with some success in children in areas with relatively high pollution.[12,13]

Measurements of pulmonary function in children have demonstrated differences especially in areas of moderate pollution. Of these tests, those involving flow such as peak expiratory flow, $FEV_{1.0}$ or volume-flow relationships have appeared to be more sensitive[10,13,14] (C. M. Shy, F. Benson, and C. J. Nelson, unpublished data).

Applications for admission to hospitals have been found to be an informative index of morbidity in London.[15] The use of such techniques is not warranted for comparative purposes between cities or between countries because criteria for hospitalization, availability of beds, medical custom, etc, differ so widely. Similar criticisms can be levelled against the use of sickness absences in industries because insurance programs, sickness benefits, and industrial medical custom vary greatly. These techniques can be applied only to a specific area where the above variables remain relatively constant. They can then be used to assess the possible effects of varying levels of pollutants or climatic changes in a prospective manner. Even so, there are a number of factors that influence admission to hospitals, such as day of week, availability of beds, etc. The minimization of these effects requires sophisticated statistical analysis.[16]

175

Objective Methods.—As with the subjective methods, standardized methods are required, and the various devices must be calibrated at regular intervals if comparisons between countries or within areas in a given country are to be made.

Attempts have been made to quantify and characterize sputum. Some of the earlier studies in Great Britain demonstrated a good correlation between volume-collected and positive responses to questions of phlegm production. Japanese studies have not shown such good correlation.[17] Gross character of the sputum tends to be variable because of the presence or absence of chest colds at the time the sample is collected. Studies in the United States have tended to confirm the British ones, but cooperation is poorer. These studies thus validate the relationship of sputum production to the responses to questions for Anglo Saxon-Caucasian populations. In general, this is not a worthwhile survey technique unless there is a specific reason for sputum collection, as for bacteriological and cytological studies; on the other hand, it is a useful check on differences identified by the questionnaire between countries.[17,18]

Nasal smears of secretions or blood samples for eosinophils or immunoglobulins have received limited study and deserve more investigation as possible epidemiological tools. Immunoglobulins may identify a sensitive or resistant portion of the population; for example, the association of a deficiency in serum alpha, antitrypsin activity, and familial emphysema. Immunoglobulin E appears to be associated with extrinsic asthma; specific blood groups have been reported to be more commonly present in those persons manifesting silicosis after different periods of exposure.[19]

Tests of Pulmonary Function.—Spirometry has been a standard method in the past. Certain tests will be used in the future to maintain comparability. In general, those tests involving flow are more sensitive than those measuring volume alone. Values should be corrected to body temperature, pressure, saturation where possible (peak flows obtained by a Wright peak-flow meter cannot be so corrected). Whether maximum or mean values are used does not appear to be important. Methods, however, must be clearly specified, eg, number of trials, which trials were used, and type of apparatus.

Volume-flow plots can be made in the field either by means of a special device[20] or spirometer tracings.[21,22] The exact type of apparatus used will be determined by availability and costs. Since the portion of the curve that is developed at the lower-lung volume (from 50% of FVC down) seems to be the most sensitive and most useful, a body plethysmograph is not needed since simpler methods can obtain adequate data.

Respiratory resistance can be measured by the oscillating technique[23]; this technique is proving to be useful in surveys. The simplest method is to use a fixed frequency (3 cycles per second) and fixed pressure. The flows so generated on top of quiet respiration permit measuring the respiratory flow resistance. Comparisons of this method with other methods of measuring pulmonary or airway resistance have shown good agreement.[24] Combining results from this technique with flows obtained on the percent-volume-in-volume relationships may also prove to be useful to estimate specific pulmonary-driving pressure.

The failure rate using these methods (spirometry, volume-flow, peak flow, or oscillating resistance) is extremely low, 1% to ½% in general population surveys and lower in occupational groups. Failure rate includes the few who refuse the test plus a larger number who are unable to perform the coordinated activity for the test.

X-ray films of the lungs (posteroanterior and lateral) are a valuable survey method for a number of reasons; they can be used to obtain total lung capacity.[25] Measurement of total lung capacity permits us to relate the measured resistance and driving pressure to an absolute lung volume. The x-ray film method is much quicker and less difficult for the subject. The film can also be read for possible pulmonary or cardiac disease. More data of this sort are indicated to obtain normal values in general populations and to determine whether this measurement is more sensitive than the other tests of pulmonary function. In certain occupational groups, such as asbestos workers, it is very useful.

The diffusing capacity of the lungs has been measured in some surveys[26] (W. F. Van Ganse and B. J. Ferris Jr., unpublished data). The single-breath method has been used most because it is easier and less complicated than the steady-state method. This measurement requires more complex equipment than the other tests and considerable cooperation on the part of the subject. Failure rates, therefore, tend to be higher than for the other tests (10% to 15%) in general population surveys and lower in occupational groups. Measuring the diffusing capacity of the lungs for carbon monoxide ($D_L CO$) on high oxygen may be worth the extra effort. If this is done, one can calculate membrane component and volume of blood in capillaries. In some surveys, it may be better to obtain two or three measurements of $D_L CO$ on room air. An estimate of the effective lung volume is also obtained from the dilution of the helium. This value can be compared with the x-ray film method to obtain an index of ventilated lung.

Carbon monoxide hemoglobin can be estimated from exhaled air.[27-28] This has relevance to cardiovascular disease in view of the recent observations of an increased mortality due to cardiovascular disease with chronic exposures to relatively low levels of CO.[29] Further studies are needed to confirm or refute these observations.

Cardiovascular.—Electrocardiograms have been used in a number of studies.[30-32] Their usefulness still needs more documentation. Data should be collected or analyzed or both to obtain a set of normal standards and the variation to be expected in several populations. Efforts are being made in this direction.[33] Their value, in conjunction with

178

an added stress such as exercise, should be assessed, and their usefulness as a predictor of future mortality or morbidity are indicated.

Measurements of blood pressure have shown considerable observer variations.[34] This may be a useful parameter to include in surveys as more information is obtained on the effects of certain metals such as cadmium and lead on the cardiovascular system. The methods of measuring the blood pressure must be standardized[34,35] and probably should be repeated after a period of rest and allaying of any anxiety in the subject. The measurement of pulse per se has limited value. If measured in conjunction with exercise or equivalent stress, and the response during exercise as well as the rate of recovery is followed, considerable information may be obtained. Such an addition in surveys is only warranted when emphasis is placed on physical fitness or exercise capability.

Blood cholesterol and blood clotting should be obtained only in selected situations where specific questions concerning these components are being asked. Alterations in blood clotting may be factors in the response to CO. Thus, an investigation of the effects of various CO-hemoglobin levels and blood clotting is indicated.

Studies in Europe have shown alterations in the red blood cells as a result of exposure to high levels of air pollution.[36] Although nutrition was not thought to be a factor in those studies, there is reason to question it. It is doubtful that there are levels of air pollution in North America where this could be tested.

The effect of pollutants on exercise capacity deserves further investigation. A report from California[37] indicates that long-distance runners had poorer times when air pollution (oxidant) was high. Some studies have been reported in which subjects have been exposed to various air pollutants during varying degrees of exercise.[38,39] The responses reported should be verified and studied in more detail.

Studies are in progress in which retinal vessels are photographed after fluorescein injections. This method is still in the developmental stage. The side effects of intravenously administered fluorescein also militate against the general use of this procedure. As the technique is improved, it may prove to be a useful method of examining the small blood vessels but only in highly selected population groups.

Hepatorenal-Intestinal.—Reports have indicated that pollutants such as those around a pulp mill using the Kraft process may cause nausea, headache, and some gastrointestinal symptoms (quoted in Anderson).[9] These have been vague. Although these organ systems may prove to be influenced in human beings, it is not feasible to test them by animal studies, as has been done with other systems.

In selected situations, urinalyses may be indicated as in possible cadmium, lead, or uranium exposures. In our present stage of the art, these areas seem to be less profitable than the others for immediate study. The reported association of increased cadmium levels in the air to cardiovascular-disease death rate[40] deserves further investigation.

Nervous System.—Russian studies have placed much emphasis on sensory and other neurological changes to low levels of pollution.[41] The significance of these is difficult to assess, but attempts should be made to verify them in this country.

Another study has indicated that relatively short exposures (90 minutes) to 50 ppm of CO can produce changes in tone-length discrimination. Pure tones were used. The second was of longer, shorter, or the same duration as the first. The percent-correct responses were plotted against CO levels and were found to decrease with increasing levels of CO. Certainly these studies should be extended.[42]

Personal Dosimetry.—Hair and nail clippings can be analyzed to obtain estimates of exposure to a variety of substances. Lead, arsenic, cadmium, mercury, and other metals can be measured with considerable accuracy at very low concentrations.[43] Analyses of serial segments of hair can be particularly useful. The individual thus can be his own personal sampler for selected materials.

This investigation was supported in part by Public Health Service research grant EC-00205.

This material was drawn from a background document prepared by the author as background information for the Task Force on Research Planning, prepared for the National Institute of Environmental Health Sciences. The report of the Task Force is an independent and collective report which is being published by the Government Printing Office under the title *Man's Health and the Environment: Some Research Needs*. It constitutes part of appendix B of that report. The original material for this background document, as well as others prepared for that report, is deposited in the National Library of Medicine, which will supply copies on request.

References

1. Logan WPD: Mortality from fog in London, January 1956. *Brit Med J* 1:722-725, 1956.
2. Schrenk HH, Heimann H, Clayton GD, et al: Air pollution in Donora, Pa. *Public Health Bull* **306:**171, 1949.
3. Boyd JT: Climate, air pollution, and mortality. *Brit J Prev Soc Med* 14:123-135, 1960.
4. Waller RE, Lawther PJ: Some observations on London fog. *Brit Med J* 2:1356-1358, 1955.
5. Spicer WS Jr, Storey PB, Morgan WK, et al: Variation in respiratory function in selected patients and its relation to air pollution. *Amer Rev Resp Dis* 86:705-712, 1962.
6. Burrows B, Kellogg AL, Buskey J: Relationship of symptoms of chronic bronchitis and emphysema to weather and air pollution. *Arch Environ Health* 16:406-413, 1968.
7. McCarroll J, Cassell EJ, Wolter DW, et al: Health and the urban environment: V. Air pollution and illness in a normal urban population. *Arch Environ Health* 14:178-184, 1967.
8. Damon a: Negro-white difference in pulmonary function (vital capacity, time vital capacity and expiratory flow rate). *Hum Biol* 38:380-393, 1966.
9. Anderson DO, Larsen AA: The incidence of illness among young children in two communities of different air quality: A pilot study. *Canad Med Assoc J* 95:893-904, 1966.
10. Ferris BG Jr: Effect of air pollution on incidence of illness in first and second graders, Berlin, N.H. *Amer Rev Resp Dis,* to be published.
11. Douglas JWB, Waller RE: Air pollution and respiratory infection in children. *Brit J Prev Soc Med* 20:1-8, 1966.
12. Lunn JE, Knowelden J, Handyside KJ: Patterns of respiratory illness in Sheffield infant school children. *Brit J Prev Soc Med* 21:7-16, 1967.

181

13. Watanabe H, Kaneko F, Murayama H, et al: Effect of air pollution on health, peak flow rate and vital capacity of primary school children. *Osaka City Inst Hyg* 36:32-37, 1964.

14. Toyama T: Air pollution and its health effects in Japan. *Arch Environ Health* 8:153-173, 1964.

15. Martin AE: Mortality and morbidity statistics and air pollution. *Proc Roy Soc Med* 57:969-975, 1964.

16. Sterling TD, Pollack SV, Phair JJ: Urban hospital morbidity and air pollution. *Arch Environ Health* 15:362-374, 1967.

17. Sakai Y, Matsuya T, Nakamura T: Some characteristic results of the respiratory survey for Japanese telephone men. Read before the annual meeting of the American Academy of Occupational Medicine, Boston, 1969.

18. Stone RW: Respiratory symptoms in telephone plant men. Read before the annual meeting of the American Academy of Occupational Medicine, Boston, 1969.

19. Vidakovic A, Andjelkovski A, Brusin A: Les Groupes Sanguins chez les Malades de Pneumoconiose. Fifteenth International Congress on Occupational Health. Vienna, H Egermann, 2:603-606, 1966.

20. Peters JM, Mead J, Van Ganse WF: A simple flow volume device for measuring ventilatory function in the field: Results on workers exposed to low levels of TDI (toluene diisocyanate). *Amer Rev Resp Dis* 99:617-622, 1969.

21. Peters JM, Ferris BG Jr: Smoking, pulmonary function and respiratory symptoms in a college-age group. *Amer Rev Resp Dis* 95:774-782, 1967.

22. Ferris BG Jr, Burgess WA, Worcester J: Prevalence of chronic respiratory disease in a pulp mill and a paper mill in the United States. *Brit J Industr Med* 24:26-37, 1967.

23. Grimby G, Takishima T, Graham W, et al: Frequency dependence of flow resistance in patients with obstructive lung disease. *J Clin Invest* 47:1455-1465, 1968.

24. Goldman M, Knudson R, Mead J, et al: A simplified measurement of respiratory resistance by forced oscillations. *J Appl Physiol*, to be published.

25. Pratt PC, Klugh GA: A method for determination of total lung capacity from posteroanterior and lateral chest roentgenograms. *Amer Rev Resp Dis* 96:548-552, 1967.

26. Bates DV, Gordon CA, Place REG, et al: Chronic bronchitis: Report on the third and fourth stages of the coordinated study of chronic bronchitis in the Department of Veterans Affairs. *Canad Med Services J* 22:5-59, 1966.

27. Jones RH, Ellicotte MF, Cadigan JB, et al: The relationship between alveolar and blood carbon monoxide concentrations during breath holding. *J Lab Clin Med* 51:553-564, 1958.

28. Ringold A, Goldsmith JR, Helwig HL, et al: Estimating recent carbon monoxide exposures. *Arch Environ Health* 5:308-317, 1962.

29. Goldsmith JR: Carbon monoxide and coro-

nary artery disease. *Ann Intern Med* **71**:199-201, 1969.

30. Rose GA: The coding of survey electrocardiograms by technicians. *Brit Heart J* **27**:595-598, 1965.

31. Higgins ITT, Kannel WB, Dawley TR: The electrocardiogram in epidemiological studies. *Brit J Prev Soc Med* **19**:53-68, 1965.

32. Burrows B, Earle RH: Course and prognosis of chronic obstructive lung disease: A prospective study of 200 patients. *New Eng J Med* **280**:397-404, 1969.

33. Kagan A: Comparability of electrocardiographic data with particular reference to the W.H.O Minnesota code trial. *Milbank Mem Fund Quart* **43**:40-48, 1965.

34. Rose GA, Holland WW, Crowley EA: A sphygmomanometer for epidemiologists. *Lancet* **1**:296-300, 1964.

35. Rose GA: Standardization of observers in blood pressure measurements. *Lancet* **1**:673-674, 1965.

36. Kapalik VL: The red blood picture in children from different environments. *Rev Czech Med* **9**:65-81, 1963.

37. Wayne WS, Wehrle PF, Carroll RE: Oxidant air pollution and athletic performances. *JAMA* **199**:901-904, 1967.

38. Smith L: Inhalation of the photochemical smog compound peroxy acetyl nitrate. *Amer J Public Health* **55**:1460-1468, 1965.

39. Holland GJ, Benson D, Bush A, et al: Air pollution simulation and human performance. *Amer J Public Health* **58**:1684-1691, 1968.

40. Carroll RE: The relationship of cadmium in the air to cardiovascular disease death rate. *JAMA* **198**:267-269, 1966.

41. Ryazanov VA: Sensory physiology as basis for air quality standards. *Arch Environ Health* **5**:480-494, 1962.

42. Beard RR, Wertheim GA: Behavioral impairment associated with small doses of carbon monoxide. *Amer J Public Health* **56**:2012-2022, 1967.

43. Kopito L, Byers RK, Schwachman H: Lead in hair of children with chronic lead poisoning. *New Eng J Med* **276**:949-953, 1967.

Air Pollution and Human Health

The quantitative effect, with an estimate of the dollar benefit of pollution abatement, is considered.

Lester B. Lave and Eugene P. Seskin

Air pollution is a problem of growing importance; public interest seems to have risen faster than the level of pollution in recent years. Presidential messages and news stories have reflected the opinion of scientists and civic leaders that pollution must be abated. This concern has manifested itself in tightened local ordinances (and, more importantly, in increased enforcement of existing ordinances), in federal legislation, and in extensive research to find ways of controlling the emission of pollutants from automobiles and smokestacks. Pollutants are natural constituents of the air. Even without man and his technology, plants, animals, and natural activity would cause some pollution. For ex-

ample, animals vent carbon dioxide, volcanic action produces sulfur oxides, and wind movement insures that there will be suspended particulates; there is no possibility of removing all pollution from the air. Instead, the problem is one of balancing the need of polluters to vent residuals against the damage suffered by society as a result of the increased pollution (1). To find an optimum level, we must know the marginal costs and marginal benefits associated with abatement. This article is focused on measuring one aspect of the benefit of pollution abatement.

Polluted air affects the health of human beings and of all animals and plants (2). It soils and deteriorates property, impairs various production processes (for example, the widespread use of "clean rooms" is an attempt to reduce contamination from the air), raises the rate of automobile and airline accidents (3), and generally makes living things less comfortable and less happy. Some of these effects are quite definite and measurable, but most are ill-defined and difficult to measure, even conceptually. Thus, scientists still disagree on the quantitative effect of pollution on animals, plants, and materials. Some estimates of the cost of the soiling and deterioration of property have been made, but the estimates are only a step beyond guesses (4). We conjecture that the major benefit of pollution abatement will be found in a general increase in human happiness or improvement in the "quality of life," rather than in one of the specific, more easily measurable categories. Nonetheless, the "hard" costs are real and at least theoretically measurable.

In this article we report an investigation of the effect of air pollution on human health; we characterize the problem of isolating health effects; we derive quantitative estimates of the effect of air pollution on various diseases and point out reasons for viewing some earlier estimates with caution; we discuss the economic costs of ill health; and we estimate the costs of effects attributed to air pollution.

The Effect of Air Pollution
on Human Health

In no area of the world is the mean annual level of air pollution high enough to cause continuous acute health problems. Emitted pollutants are diluted in the atmosphere and swept away by winds, except during an inversion; then. for a period that varies from a few hours to a week or more, pollutants are trapped and the dilution process is impeded. When an inversion persists for a week or more, pollution increases substantially, and there is an accompanying increase in the death rate.

Much time has been spent in investigating short-term episodes of air pollution (5). We are more concerned with the long-term effects of growing up in, and living in, a polluted atmosphere. Few scientists would be surprised to find that air pollution is associated with respiratory diseases of many sorts, including lung cancer and emphysema. A number of studies have established a qualitative link between air pollution and ill health.

A qualitative link, however, is of little use. To estimate the benefit of pollution abatement, we must know how the incidence of a disease varies with the level of pollution. The number of studies that allow one to infer a quantitative association is much smaller.

Quantifying the relation. Our objective is to determine the amount of morbidity and mortality for specific diseases that can be ascribed to air pollution. The state of one's health depends on factors (both present and past) such as inherited characteristics (that cause a predisposition to certain diseases), personal habits such as smoking and exercise, general physical condition, diet (including the amount of pollutants ingested with food), living conditions, urban and occupational air pollution, and water pollution (6, 7). Health is a complex matter, and it is exceedingly difficult to sort out the contributions of the various factors. In trying to determine the contribution of any single factor one must be careful neither to include spurious effects nor to conclude on the basis of a single insignificant correlation that there is no association. Laboratory experimentation is of little help in the sorting process (8).

The model implicit in the studies we have examined is a simple linear equation wherein the mortality or morbidity rate is a linear function of the measured level of pollution and, possibly, of an additional socioeconomic variable. In only a few cases do the investigators go beyond calculating a simple or partial correlation.

A number of criticisms can be leveled at ths simple model. No account is taken of possibly important factors such as occupational exposure to air pollution and personal habits. These and other factors influencing health must be uncorrelated with the level of pollution, if the estimated effect of pollution is to be an unbiased estimate. In addition, the linear form of the function is not very plausible, except insofar as one considers it a linear approximation over a small range.

Both because of the rather crude nature of the studies and because of the statistical estimation, there is a range of uncertainty concerning the quantitative effect of pollution on human health. This range is reflected in the estimate of the benefit of pollution abatement, discussed below.

Epidemiological studies. Epidemiological data are the kind of health data best adapted to the estimation of air pollution effects. These data are in the form of mortality (or morbidity) rates for a particular group, generally defined geographically (*9*). For example, an analyst may try to account for variations in the mortality rate among the various census tracts in a city. While these vital statistics are tabulated by the government and so are easily available, there are problems with the accuracy of the classification of the cause of death (since few diagnoses are verified by autopsy and not all physicians take equal care in finding the cause of death). Other problems stem from unmeasured variables such as smoking habits, occupations, occupational exposure to air pollution, and genetic health factors. Whenever a variable is unmeasured, the analyst is implicitly assuming either that it is constant across groups or else that it varies randomly with respect to the level of air pollution. Since there are many unmeasured variables, one should not be surprised to discover that some studies fail to find a significant relationship or that

others find a spurious one. For the same reason, one should not expect the quantitative effect to be identical across various groups, even when. the relationship in each group is statistically significant.

Sample surveys are a means of gathering a more complete set of data. For example, a retrospective analysis might begin with a sample of people who died from a particular disease. Through questionnaires and interviews, the smoking habits and residence patterns of the deceased can be established. The analysis would then consist of an attempt to find the factors implicated in the death of these individuals. Two types of problems arising from such a study are the proper measurement of variables such as exposure to air pollution (there are many pollutants and many patterns of lifetime exposure) and the possible contributions of variables which still are unobserved, such as occupational hazards, socioeconomic characteristics, and personal habits.

Whatever the source of data, the investigators must rest their cases by concluding that the associations which they find are so strong that it is extremely unlikely that omitted variables could have given rise to the observed correlations; they cannot account for all possible variables.

Episodic relationships. Another method of investigating the effects of pollution involves an attempt to relate daily or weekly mortality (or morbidity) rates to indices of air pollution during the interval in question (*10*). The conclusions of these studies are of limited interest, for two reasons. First, someone who is killed by an increase in air pollution is likely to be gravely ill. Air pollution is a rather subtle irritant, and it is unlikely that a healthy 25-year-old will succumb to a rise in pollution levels. Our interest should be focused on the initial cause of illness rather than on the factor that is the immediate determinant of death. Thus, morbidity data are more useful than mortality data. Second (and more important for the morbidity studies), there are many factors that affect the daily morbidity rate or daily rate of employee absences. Absence rates tend to be high on Mondays and Fridays for reasons that have nothing to do with air pollution or illness. One would expect little change in these absence rates if air pollution were reduced. Other factors, such as absence around holidays, give rise to spurious variation; this can be handled by ignoring the periods in question or by gathering enough data so that this spurious variation is averaged away. Some of these factors (such as high absence rates on Fridays and seasonal absence rates) may be correlated with variations in air pollution and no amount of data or of averaging will separate the effects. We have chosen to disregard the results of these episodic studies, with a few exceptions, cited below.

It is difficult to isolate the pollutants that have the most important effects on health on the basis of the studies we survey here. Measurement techniques have been crude, and there has been a tendency to base concentration figures on a single measurement for a large area. A more important problem is the fact that in most of these studies only a single pollutant was reported. Discovering which pollutants are most harmful is an important area, where further exploration is necessary. We have tried, nevertheless, to differentiate among pollutants in the survey that follows (11). The problem is complicated, since pollution has increased over time, and since lifetime exposure might bear little relation to currently measured levels. These problems are discussed elsewhere (12).

A Review of the Literature

We will proceed with a detailed review of studies made in an attempt to find an association between mortality or morbidity and air pollution indices.

Air pollution and bronchitis. Studies link morbidity and mortality from bronchitis to air pollution in England (13), the United States (14, 15), Japan (16), and other countries (17). Mortality rates by country boroughs in England and Wales have been correlated with pollution (as measured by the sulfation rate, total concentration of solids in the air, a deposit index, and the density of suspended particulates) and with socioeconomic variables (such as population density and social class). The smoking habits of the individuals studied have also been investigated. The conclusion of these studies is that air pollution accounts for a doubling of the bronchitis mortality rate for urban, as compared to rural, areas.

We took data reported by Stocks (18, 19) and by Ashley (20) and performed a multiple regression analysis, as shown in Table 1. We fit the following equation to the data

$$MR_i = a_0 + a_1 P_i + a_2 S_i + e_i \quad (1)$$

where MR_i is the mortality rate for a particular disease in country borough i, P_i is a measure of air pollution in that borough, S_i is a measure of socioeconomic status in borough i, and e_i is an error term with a mean of zero. (We also fit other functional forms, as discussed below.) Under general assumptions, the estimated coefficients (a_0, a_1, a_2) will be best linear, unbiased estimates (21). Only if we want to perform significance tests must we make an assumption about the distribution of the error term (for example, the assumption that it is distributed normally).

The first regression in Table 1 relates the bronchitis mortality rate for men to a deposit index (see Table 1, footnote †), and the population density in each of 53 country boroughs. Thirty-nine percent of the variation in the mortality rate (across boroughs) is "explained" by the regression. It is estimated that a unit increase in the deposit index (1 gram per 100 square meters per month) leads to an increase of 0.18 percent in the bronchitis mortality rate (with popu-

Table 1. Multiple regressions based on data from England.¹ Numbers in parentheses are the t statistic.

Category	R^2*	Air pollution	Socio-economic
Bronchitis mortality rate			
1. Males, 53 county boroughs‡ (deposit index, persons/acre)	.386	.182 (4.80)	.016 (.22)
2. Females	.332	.182 (4.55)	−.031 (−.42)
3. Males, 28 county boroughs§ (smoke, persons/acre)	.433	1.891 (3.79)	.180 (1.86)
4. Females	.412	1.756 (3.23)	.252 (2.40)
5. Males, 26 areas‖ (smoke, persons/acre)	.766	.310 (3.77)	.062 (.53)
6. Females	.559	.303 (2.85)	−.038 (−.25)
7. Males, 26 areas (smoke, social class)	.783	.301 (5.86)	.176 (1.44)
8. Females	.601	.213 (3.31)	.248 (1.59)
9. Both sexes, 53 urban areas ¶ (smoke, persons/acre)	.377	.199 (4.07)	.159 (3.02)
10. Both sexes, 53 urban areas (SO₂ persons/acre)	.300	.161 (3.05)	.151 (2.64)
Lung cancer mortality rate			
11. 53 County boroughs (deposit index, persons/acre)	.445	.041 (2.09)	.154 (4.23)
12. 28 County boroughs (smoke, persons/acre)	.576	.864 (4.08)	.161 (3.89)
13. Male, 26 areas (smoke, persons/acre)	.781	.137 (2.86)	.115 (1.70)
14. Male, 26 areas (smoke, social class)	.805	.161 (5.62)	.172 (2.47)
15. 53 Urban areas (smoke, persons/acre)	.344	−.086 (−2.42)	.184 (4.83)
16. 53 Urban areas (SO₂, persons/acre)	.378	−.105 (−3.00)	.197 (5.23)
Other cancers			
17. Stomach, male, 53 county boroughs (deposit index, persons/acre)	.167	.070 (3.08)	.005 (.12)
18. Stomach, female	.175	.070 (3.08)	−.023 (−.56)
19. Stomach, male, 28 county boroughs (smoke, persons/acre)	.257	.714 (2.57)	.065 (1.21)
20. Stomach, female	.454	.883 (4.13)	.066 (1.60)
21. Intestinal, 53 county boroughs (deposit index, persons/acre)	.041	.018 (1.45)	−.012 (−.52)
22. Intestinal, 28 county boroughs (smoke, persons/acre)	.129	.174 (1.26)	.036 (1.35)
23. Other cancer, male, 26 areas (smoke, persons/acre)	.454	.019 (.59)	.073 (1.60)
24. Other cancer, female, 26 areas (smoke, persons/acre)	.044	.039 (.93)	−.062 (−1.03)
25. Other cancer, male, 26 areas (smoke, social class)	.396	.060 (2.75)	.017 (.33)
26. Other cancer, female, 26 areas (smoke, social class)	.002	.005 (.17)	−.013 (−.19)
Pneumonia mortality rate			
27. Male, 26 areas (smoke, persons/acre)	.477	.118 (1.34)	.121 (.97)
28. Female	.253	.068 (.58)	.137 (.83)
29. Male, 26 areas (smoke, social class)	.475	.158 (2.82)	.126 (.93)
30. Female	.242	.124 (1.65)	.106 (.58)

* The coefficient of determination: a value of .386 indicates a multiple correlation coefficient of .62, and indicates that 39 percent of the variation in the death rate is "explained" by the regression. † The t statistic: for a one-tailed t-test with 23 degrees of freedom, a value of 1.71 indicates significance at the .05 level; for 25 or 50 degrees of freedom, the critical values are 1.71 and 1.68. ‡ Data for 53 county boroughs in England and Wales as reported by Stocks (18). Air pollution is measured by a deposit index (in grams per 100 square meters per month) whose observed range is 96 to 731, with a mean of 375. The socioeconomic index is expressed in numbers of persons per acre (multiplied by 10); the range is 69 to 364, and the mean is 163. Death rates are measured as index numbers, with the mean for all boroughs in England and Wales equal to 100. Ranges within this sample are as follows: bronchitis (males), 73 to 259; bronchitis (females), 72 to 268; lung cancer, 70 to 159; stomach cancer (males), 67 to 168; stomach cancer (females), 84 to 161; intestinal cancer, 87 to 123. § Data for 28 county boroughs in England and Wales as reported by Stocks (18). Air pollution is measured by a smoke index (suspended matter, in milligrams per 100 cubic meters); the range is 6 to 49. Again, the socioeconomic index is expressed in numbers of persons per acre ($\times$ 10); the range is 83 to 342. ‖ Data for 26 areas in northern England and Wales as reported in Stocks (19). Air pollution is measured by a smoke index, as for category 3; the range is 15 to 562 mg/1000 m^3 and the mean is 260. One socioeconomic variable is the number of persons per acre ($\times$ 10); the range is 1 to 342 and the mean is 102. The other socioeconomic variable is social class; the range is 61 to 295. Death rates are measured as for category 1; within this sample, the range for lung cancer is 23 to 165; for other cancer, 6 to 122 (males) and 88 to 154 (females); for bronchitis, 18 to 259 (males) and 12 to 240 (females); for pneumonia, 61 to 227 (males) and 40 to 245 (females). ¶ Data for 53 areas as reported by Ashley (20). Air pollution is measured (i) by a smoke index (as for category 3), with a range of 23 to 261 μg/m^3 and a mean of 124, or (ii) by an SO_2 index (apparently in the same units), with a range of 33 to 277 and a mean of 124. Death rates are measured as for category 1; within this sample, the range for lung cancer is 70 to 146, and for bronchitis, 64 to 186.

lation per acre held constant). An increase of 0.1 person per acre in the population density is estimated to lead to an increase of 0.02 percent in the mortality rate (with air pollution held constant). As indicated in Table 1 by the t statistics (the values in parentheses below the estimated coefficients), the air pollution variable is extremely important, whereas the socioeconomic variable contributes nothing to the explanatory power of the regression.

The first ten regressions in Table 1 are an attempt to explain the bronchitis death rate. Four different data sets are used, along with three measures of pollution and two socioeconomic variables. The coefficient of determination, R^2 (the proportion of the variation in the mortality rate explained by the regression), ranges from .3 to .8. Air pollution is a significant explanatory variable in all cases. In only three cases is the socioeconomic variable significant.

The implication of the first regression is that a 10 percent decrease in the deposit rate (38 g 100 m^{-2} month $^{-1}$) would lead to a 7 percent decrease in the bronchitis death rate. Another way of illustrating the effect of air pollution on health is to note that, if all the boroughs were to improve the quality of their air to that enjoyed by the borough having the best air of all those in the sample (a standard deposit rate for all boroughs of 96 g 100 m^{-2} month $^{-1}$), the average mortality rate (for this sample) would fall from 129 to 77. Thus, cleaning the air to the level of cleanliness enjoyed by the area with the best air would mean a 40 percent drop in the bronchitis death rate among males. In the fifth regression the pollution index is a smoke index (Table 1, footnote §), and a different set of areas is considered. This is a more successful regression in terms of the percentage of variation explained. As before, the air pollution coefficient is extremely significant, and the implication is that cleaning the air to the level of cleanliness currently enjoyed by the area with

the best air (15 mg/100 m³) would lower the average bronchitis mortality rate from 106 to 30, a drop of 70 percent. Results of the other regression analyses based on bronchitis mortality data have similar implications. Note that the effect is almost the same for males and females. This indicates reliability and suggests that the effect is independent of occupational exposure.

Winkelstein *et al.* (*14*) collected data on 21 areas in and around Buffalo, New York. A cross tabulation of census tracts by income level and pollution level shows that the mortality rate for asthma, bronchitis, and emphysema (in white males 50 to 69 years old) increases by more than 100 percent as pollution rises from level 1 to level 4 (see *22*).

These studies indicate a strong relationship between bronchitis mortality and a number of indices of air pollution. We conclude that bronchitis mortality could be reduced by from 25 to 50 percent depending on the particular location and deposit index, by reducing pollution to the lowest level currently prevailing in these regions. For example, if the air in all of Buffalo were made as clean as the air in those parts of the area that have the best air, a reduction of approximately 50 percent in bronchitis mortality would probably result.

Air pollution and lung cancer. The rate of death from lung cancer has been correlated with several indices of pollution and socioeconomic variables in studies that provided controls for smoking habits and other factors. For English nonsmokers, Stocks and Campbell (*23, 24*) found a tenfold difference between the death rates for rural and urban areas. Daly (*25*), in comparing death rates in urban and rural areas of England and Wales, found the urban rate twice as high. Evidence for other parts of Europe also shows an association between lung cancer and air pollution (*26*).

Regressions 11 through 16 (Table 1) show our reworking of the data for lung cancer mortality for England and Wales (there is no control for smoking). Regressions 11 through 14 imply that, if the quality of air of all boroughs were improved to that of the borough with the best air, the rate of death from lung cancer would fall by between 11 and 44 percent. Regressions 15 and 16 show a relationship between air pollution and lung cancer which is either insignificant or inverse. The only contrary results come from Ashley's data. In the absence of more complete evidence, we must remain curious about these results. Use of such small samples and inadequate controls is certain to lead to some contrary results, but they are disconcerting when they appear.

In a study of 187,783 white American males (50 to 69 years old), Hammond and Horn (27) reported that the age-standardized rate of death due to lung cancer was 34 (per 100,000) in rural areas as compared to 56 in cities of population over 50,000. When standardized with respect both to smoking habits and to age, the rate was 39 in rural areas and 52 in cities of over 50,000.

Haenszel et al. (28) analyzed 2191 lung cancer deaths among white American males, that had occurred in 46 states, and data for a control group consisting of males who died from other causes. They found the crude rate of death from lung cancer to have been 1.56 times as high in the urban areas of their study as in the rural areas in 1958 and 1.82 times as high in the period 1948–49 (in subjects 35 years and older, with adjustments made for age). When adjustments are made for both age and smoking history, the ratio is 1.43. Also the ratio increased with duration of residence in the urban or rural area, from 1.08 for residence of less than 1 year to 2.00 for lifetime residence. Haenszel and Taeuber (29) report similar results for white American females. In a number of additional studies the association between air pollution and lung cancer is examined (30).

Buell and Dunn (31) review the evidence on lung cancer and air pollution; a summary of their findings is given in Table 2. For smokers, death rates (adjusted for age and smoking) ranged from 25 to 123 percent higher in urban areas than in rural areas. For nonsmokers, all differences exceeded 120 percent. "The etiological roles for lung cancer of urban living and cigarette smoking seem each to be complete," they say, "in that the urban factor is evident when viewing nonsmokers exclusively, and the smoking factor is evident when viewing rural dwellers exclusively." They argue that differences in the quality of diagnosis could not account for the observed differences for urban and rural areas.

Nonrespiratory-tract cancers and air pollution. Our reworking of data from England on rates of death from nonrespiratory-tract cancer is presented in Table 1 (regressions 17 through 26). In the regressions, stomach cancer is significantly related to a deposit index and a smoke index. The effects are nearly identical for males and females. Intestinal cancer appears to be only marginally related to indices of either deposit or smoke. For 26 areas in northern England and Wales, there appears to be little relationship between nonrespiratory-tract cancers and a smoke index. The single exception in the four regressions occurs for males when the socioeconomic variable is social class; here the smoke index explains a significant amount of the variation in the cancer mortality rate. (Apparently population density and smoke index are so highly related in these 26 areas that neither has significant power to explain such variation.)

Winkelstein and Kantor (32) investigated rates of mortality from stomach cancer in Buffalo, New York, and the immediate environs. Their measure of pollution is an index of suspended particulates averaged over a 2-year period. They found the rate of mortality due to stomach cancer to be more than twice

191

as great in areas of high pollution as in areas of low pollution (33).

Hagstrom et al. (34) tabulated rates of death from cancer among middle class residents of Nashville, Tennessee, between 1949 and 1960, using four measures of air pollution. They found the cancer mortality rate to be 25 percent higher in polluted areas than in areas of relatively clean air (35). They also found significant mortality-rate increases associated with individual categories of cancer, such as stomach cancer, cancer of the esophagus, and cancer of the bladder. The individual mortality rates are more closely related to air pollution after the data are broken down by sex and race.

Levin et al. (36) report, for all types of cancer, these relationships: The age-adjusted cancer-incidence rates for urban males was 24 percent higher than that for rural males in New York State (exclusive of New York City) (1949–51), 36 percent higher in Connecticut (1947–51), and 40 percent higher in Iowa (1950); the incidence rate for urban females was 14 percent higher than that for rural females in New York State, 28 percent higher in Connecticut, and 34 percent higher in Iowa. For both males and females, the incidence rate for each of 16 categories of cancer was higher in urban than in rural areas.

Cardiovascular disease and air pollution. Enterline et al. (37) found that mortality from heart disease is higher in central-city counties than in suburban counties, and, in turn, higher in suburban counties than in nonmetropolitan counties. Zeidberg et al. (38) found that both morbidity and mortality rates for heart disease are associated with air pollution levels in Nashville. The morbidity rate was about twice as high in areas of polluted air as in areas of clean air. The mortality rate was less closely associated; it was 10 to 20 percent higher in areas of polluted air than in areas of clean air (39).

Friedman (40) correlated the rate of mortality from coronary heart disease in white males aged 45 to 64 with the proportion of this group living in urban areas. The simple correlation for 33 states is .79. When cigarette consumption is held constant, the partial correlation is .67.

On the basis of these studies we conclude that a substantial abatement of air pollution would lead to 10 to 15 percent reduction in the mortality and morbidity rates for heart disease. We caution the reader that the evidence relating cardiovascular disease to air pollution is less comprehensive than that linking bronchitis and lung cancer to air pollution.

Total respiratory disease (41). Daly (25) found significant correlations between air pollution and death rates for all respiratory diseases (and for non-respiratory diseases as well) in England. Douglas and Waller (42) found significant relationships between air pollution and respiratory disease in 3866 British school children. Fairbairn and Reid (43) found significant correlations between air pollution and morbidity rates (for bronchitis, pneumonia, pulmonary tuberculosis, and lung cancer) in England. Regressions 27 through 30 in Table 1 show pneumonia mortality to be related only marginally to a smoke index.

Zeidberg et al. (44) questioned 9313 Nashville residents about recent illnesses. Among males aged 55 and older from white middle-class families, the numbers of illnesses per respondent during the past year were 1.92, 1.15, and 1.26 for areas of high, moderate, and low pollution, respectively. There are a number of other comparisons, based on other measures of air pollution and on data for females and nonwhites (some of these are given in 45). However, we should add a word of caution: although the sample size in this study was large and controls for many socioeconomic variables were included, many important factors were ignored—for example, smoking habits and length of residence. Nonetheless, the finding is extremely strong and seems unlikely to

Table 2. A summary of lung cancer mortality studies. Number of deaths from lung cancer per 100,000 population [from Buell and Dunn (31)].

Standardized for age and smoking			Nonsmokers			Study
Urban	Rural	Urban/ rural	Urban	Rural	Urban/ rural	
101	80	1.26	36	11	3.27	Buell, Dunn, and Breslow* (67)
52	39	1.33	15	0	∞	Hammond and Horn (68)†
189	85	2.23	50	22	2.27	Stocks (69)‡
			38	10	3.80	Dean (70)§
149	69	2.15	23	29	.79	Golledge and Wicken (71) ‖
100	50	2.00	16	5	3.20	Haenszel et al. (72). ¶

* California men; death rates by counties.　† American men.　‡ England and Wales.　§ Northern Ireland.　‖ England; no adjustment for smoking.　¶ American men.

be an artifact of unmeasured variables.

Hammond (46) studied over 50,000 men to find the relationships between emphysema, age, occupational exposure to pollution, urban exposure, and smoking. His results indicated that the effect of air pollution is significant and that heavy smokers have a much higher morbidity rate in cities than in rural areas; the effect becomes more marked as age increases.

Ishikawa et al. (47) estimated the incidence of emphysema in Winnipeg (Canada) and St. Louis. They examined the lungs of 300 corpses in each city (the samples were comparable). Findings for each age group (over 25 years old) indicated that the incidence and severity of emphysema is higher in St. Louis, the city with the more polluted air. (In the 45-year-old group 5 percent of those in Winnipeg and 46 percent of those in St. Louis showed evidence of emphysema.)

A number of studies have been made in England on homogeneous occupational groups, such as postmen. The results are relatively pure in that all members of the sample have comparable incomes, working conditions, and social status. Holland and Reid (48) found that the rates of occurrence of severe respiratory symptoms were 25 to 50 percent higher for London postmen than for small-town postmen (sample size, 770). Reid (49) found that, in the postmen of his study, absences due

to bronchitis rose from an index number of 100 for the area of least air pollution, to 120 for an area of moderate pollution, to 250 and 283 for the areas of highest pollution. Corresponding figures for absences due to other respiratory illness were 100, 100, 150, and 151, respectively, and for absences due to infectious and parasitic diseases, 100, 115, 130, and 140. Cornwall and Raffle (50) made a similar study of bus drivers in London. They found that 20 to 35 percent of absences due to sickness of any kind could be ascribed to air pollution (they used a fog index as a measure of pollution). Fairbairn and Reid (43) tabulated absences due to sickness for postmen, for males working indoors, and for females working indoors. They found that the age-standardized morbidity rate for bronchitis and pneumonia in the postmen of their study rose from 40 man-years, per 1000 man-years, for the area of lowest air pollution (of the four areas studied) to 122 for the area of highest air pollution. Corresponding figures for morbidity from colds were 75 and 171 man-years, and for morbidity from influenza, 131 and 184 man-years. For males working indoors, the low and high morbidity rates were as follows: bronchitis and pneumonia, 32 and 39; colds, 53 and 64; influenza, 88 and 102.

Dohan (51) studied absences (of more than 7 days) of female employees

Table 3. Regressions relating infant and total mortality rates for 114 Standard Metropolitan Statistical Areas in the United States to air pollution and other factors. Values in parentheses are the t statistic.* For means and standard deviations (S.D.) of the variables, see †.

Category	R^2‡	Air pollution (minimum) concentrations)	Socioeconomic		
			P/m²§	Non-white (%)	Over 65 (%)	Poor (%)

Category	R^2‡	Air pollution (minimum) concentrations)	P/m²§	Non-white (%)	Over 65 (%)	Poor (%)
Total death rate						
1. Particulates	.804	0.102 (2.83)	0.001 (2.58)	0.032 (3.41)	0.682 (18.37)	0.013 (0.93)
2. Sulfates	.813	0.085 (3.73)	0.001 (1.86)	0.033 (3.56)	0.652 (17.60)	0.006 (0.49)
Death rate for infants of less than 1 year						
3. Particulates	.545	0.393 (3.07)		0.190 (6.63)		0.150 (3.28)
4. Sulfates	.522	0.150 (1.91)		0.200 (6.83)		0.123 (2.70)
Death rate for infants less than 28 days old						
5. Particulates	.260	0.273 (2.48)		0.089 (3.61)		0.063 (1.60)
6. Sulfates	.263	0.170 (2.57)		0.097 (3.96)		0.047 (1.23)
Fetal death rate						
7. Particulates	.434	0.274 (2.02)	0.004 (2.01)	0.171 (5.70)		0.106 (2.11)
8. Sulfates	.434	0.171 (1.95)	0.004 (1.82)	0.181 (5.87)		0.085 (1.71)

* The t statistic: for a one-tailed t-test, a value of 1.65 indicates significance at the .05 level.   † Total death rate per 10,000: mean, 91.5; S.D., 15.2. Infant death rate (age, < 1 year) per 10,000 live births: mean, 255.1; S.D., 36.1. Infant death rate (age, < 28 days) per 10,000 live births: mean, 188.0; S.D., 24.4. Fetal death rate per 10,000 live births: mean, 153.9; S.D., 34.4. Suspended particulates ($\mu g/m^3$), minimum reading for a biweekly period: mean, 45.2; S.D., 18.7. Total sulfates ($\mu g/m^3$) ($\times$ 10), minimum reading for a biweekly period: mean, 46.9; S.D., 30.6. Persons per square mile: mean, 763.4; S.D., 1387.9. Percentage of nonwhites in population ($\times$ 10): mean, 125.2; S.D., 102.8. Percentage of population over 65 ($\times$ 10): mean, 84.2; S.D., 21.2. Percentage of families with incomes under $3000 ($\times$ 10): mean, 181.6; S.D., 65.7.   ‡ The coefficient of determination: a value of .804 indicates a multiple correlation coefficient of .90, and indicates that 80 percent of the variation in the death rate is "explained" by the regression.   § Persons per square mile.

in eight Radio Corporation of America plants. He found a correlation of .96 between atmospheric concentrations of SO_3 and absences due to respiratory disease in the five cities for which complete data were available. During Asian flu epidemics there was a 200 percent increase in illness in cities with polluted air and only a 20 percent increase in those with relatively unpolluted air.

Infant mortality and total mortality rates. Sprague and Hagstrom (*52*) compared air-pollution data for Nashville with fetal and infant mortality rates for Nashville as given in census tracts (for 1955 through 1960). Controls for socioeconomic factors were not included. For infant death rates (ages 28 days to 11 months), the highest correlation was with atmospheric concentrations of SO_3 (in milligrams per 100 square centimeters per day) and was .70. For the neonatal death rates (ages 1 day to 27 days), the highest correlation was with dustfall and was .49. For infants dying during their first day whose death certificate includes mention of immaturity, the highest correlation was with dustfall and was .45. The correlation of the fetal death rate with dustfall was .58.

In a study just being completed (*53*), we have collected data for 114 Standard Metropolitan Statistical Areas in the United States and have attempted to relate total death rates and infant mortality rates to air pollution and other factors. Socioeconomic data, death rates, and air-pollution data were taken from U.S. government publications (54). Regression 1 (Table 3) shows how the total death rate in 1960 varies with air pollution levels and with socioeconomic factors. As the (biweekly) minimum level of suspended particulates increases, the death rate rises significantly. Moreover, the death rate increases with (i) the density of population of the area, (ii) the proportion of nonwhites, (iii) the proportion of people over age 65, and (iv) the proportion of poor families. Eighty percent of the variation in the death rate across these 114 statistical areas is explained by the regression.

Regression 3 shows how the 1960 infant death rate (age, less than 1 year) varies. A smaller proportion (55 percent) of the variation in the death rate is explained by the regression, although the minimum air-pollution level, the percentage of nonwhites, and the proportion of poor families continue to be significant explanatory variables. Regression 5 is an attempt to explain variation in the neonatal death rate. The results are quite similar to those of regression 3. The fetal death rate is examined in regression 7. Here the minimum air-pollution level, population density, the percentage of nonwhites, and the percentage of poor families are all significant explanatory variables.

Regressions 2, 4, 6, and 8 are an attempt to relate these death rates to the atmospheric concentrations of sulfates for the 114 statistical areas of the study. Regression 2 shows that the total death rate is significantly related to the minimum level of sulfate pollution, to population density, and to the percentage of people over age 65: 81 percent of the variation is explained. Regressions 4, 6, and 8 show that the minimum atmospheric concentration of sulfates is a significant explanatory variable in three categories of infant death rates.

One might put these results in per-spective by noting estimates on how small decreases in the air-pollution level affect the various death rates. A 10 percent decrease in the minimum concentration of measured particulates would decrease the total death rate by 0.5 percent, the infant death rate by 0.7 percent, the neonatal death rate by 0.6 percent, and the fetal death rate by 0.9 percent. Note that a 10 percent decrease in the percentage of poor families would decrease the total death rate by 0.2 percent and the fetal death rate by 2 percent. A 10 percent decrease in the minimum concentration of sulfates would decrease the total death rate by 0.4 percent, the infant mortality rate by 0.3 percent, the neonatal death rate by 0.4 percent, and the fetal death rate by 0.5 percent.

Each of the relations in Tables 1 and 3 was estimated in alternative ways, including transformation into logarithms, a general quadratic, and a "piecewise" linear form as documented elsewhere (12). The implications about the roles of air pollution and of the socioeconomic variables were unchanged by use of the different functional forms. Another result to be stressed is that, in Table 1, comparable regressions for males and females show almost precisely the same effects for air pollution. This suggests that occupational exposure does not affect these results; the result lends credence to the estimates. A result that we document elsewhere (53) is that it is the minimum level of air pollution that is important, not the occasional peaks. People dealing with this problem should worry about abating air pollution at all times, instead of confining their concern to increased pollution during inversions.

Some Caveats

In preceding sections we have described a number of studies which quantify the relationship between air pollution and both morbidity and mortality. Is the evidence conclusive? Is it possible for a reasonable man still to object that there is no evidence of a substantial quantitative association? We believe that there is conclusive evidence of such association (55).

In the studies discussed, a number of countries are considered, and differences in morbidity and mortality rates among different geographical areas, among people within an occupational group, and among children are examined. Various methods are used, ranging from individual medical examinations and interviews to questionnaires and tabulations of existing data. While individual studies may be attacked on the grounds that none manages to provide controls for all causes of ill health, the number of studies and the variety of approaches are persuasive. It is difficult to imagine how factors such as general habits, inherited characteristics, and lifetime exercise patterns could be taken into account.

To discredit the results, a critic would have to argue that the relationships found by the investigators are spurious because the level of air pollution is correlated with a third factor, which is the "real" cause of ill health. For example, many studies do not take into account smoking habits, occupational exposure, and the general pace of life. Perhaps city dwellers smoke more, get less exercise, tend to be more overweight, and generally live a more strained, tense life than rural dwellers. If so, morbidity and mortality rates would be higher for city dwellers, yet air pollution would be irrelevant. This explanation cannot account for the relationships found.

Apparently there is little systematic relationship between relevant "third" factors and the level of air pollution. An English study (19) in which smoking habits are examined reveals little evidence of differences by residence. There is evidence in the United States that smoking is more prevalent among lower socioeconomic groups (56) but income or other socioeconomic variables would account for this effect and still leave the pollution coefficient unbiased. More importantly, the correlations between air pollution and mortality are better when one is comparing areas within a city (where more factors are held constant) than when one is comparing rural and urban areas (57). It is especially hard to believe that the apparent relation between air pollution and ill health is spurious when significant effects are found in studies comparing individuals within strictly defined occupational groups, such as postmen or bus drivers (where incomes and working conditions are comparable and unmeasured habits are likely to be similar).

When there are uncontrolled factors, some studies may show inconclusive or even negative results; only by collecting samples large enough to "average away" spurious effects can dependable results be guaranteed. In the main, each of the studies cited above was based on a substantial sample. It is the body of studies as a whole that we find persuasive.

An examination of contrary results. Uncontrolled factors, together with small samples, are certain to lead to some results contrary to the weight of evidence and to our expectations. For example, in some studies (58) no attempt is made to control even for income or social status. From the evidence of studies which did provide such controls, we know that failure to control for income leads to biased results, and so we place little credence in either the positive or negative findings of studies lacking these controls.

Sampling error can be extremely important. For example, Zeidberg et al. (59) find mixed results in cross-tabulating respiratory disease mortality with level of air pollution and with income class. In general the relationships are in the expected direction, but they are often insignificant. Insignificant results might occur often, if the samples are small, even if air pollution is extremely significant, since sampling errors dominate the explanatory variables.

Another study in which sampling error is important is reported by Ferris and Whittenberger (7). They compared individuals in Berlin, New Hampshire, with residents of Chilliwack (British Columbia), Canada, and—not surprisingly in view of the small samples—failed to find significant differences in the occurrence of respiratory disease. Prindle *et al.* (60) compared two Pennsylvania towns in the same fashion. These two studies are admirable in that individuals were subjected to careful medical examinations. However, only a few hundred individuals were studied, and this means that sampling errors tend to obscure the effects of air pollution. Moreover, there were no controls for other factors, such as smoking. Also, one must be careful to control for a host of other variables if the sample is small. For example, the ethnic origins of the population and their general habits and occupations are known to affect mortality rate. It is exceedingly difficult to control for these factors; use of carefully constructed large samples seems the best answer. Finally, air pollution is measured currently, and it is generally assumed that relative levels have been constant over time and that people have lived at their present addresses for a long period. It is hardly surprising that statistical significance is not always obtained when such assumptions are necessary.

Since investigators are more reluctant to publish negative results than positive ones, and since it is more difficult to get negative results published, it is probable that we are unaware of other studies that fail to find a strong association between air pollution and ill health. We are somewhat reluctant to come to strong conclusions without knowledge of such negative results. However, there seems to be no reasonable alternative to evaluating the evidence at hand and allowing for uncertainty. Thus, we conclude that an objective observer would have to agree that there is an important association between air pollution and various morbidity and mortality rates.

The Economic Costs of Disease

Having found a quantitative association between air pollution and both morbidity and mortality, the next question is that of translating the increased sickness and death into dollar units. The relevant question is, How much is society willing to spend to improve health (to lower the incidence of disease)? In other words, how much is it worth to society to relieve painful symptoms, increase the level of comfort of sufferers, prevent disability, and prolong life? It has become common practice to estimate what society is willing to pay by totaling the amount that is spent on medical care and the value of earnings "forgone" as a result of the disability or death (61). This cost seems a vast underestimate for the United States in the late 1960's. Society seems willing to spend substantial sums to prolong life or relieve pain. For example, someone with kidney failure can be kept alive by renal dialysis at a cost of $15,000 to $25,000 per year; this sum is substantially in excess of forgone earnings, but today many kidney patients receive this treatment. Another example is leukemia in children; enormous sums are spent to prolong life for a few months, with no economic benefit to society. If ways could be found to keep patients with chronic bronchitis alive and active longer, it seems likely that people would be willing to spend sums substantially greater than the foregone earnings of those helped. So far as preventing disease is concerned, society is willing to spend considerable sums for public health programs such as chest x-rays, inoculation, fluoridation, pure water, and garbage disposal and for private health care programs such as annual physical checkups.

While we believe that the value of earnings forgone as a result of morbidity and mortality provides a gross underestimate of the amount society is willing to pay to lessen pain and premature death caused by disease, we have no other way of deriving numerical estimates of the dollar value of air-pollution abatement. Thus, we proceed with a conventional benefit calculation, using these forgone earnings despite our reservations.

Direct and indirect costs. Our figures for the cost of disease are based on *Estimating the Cost of Illness,* by Dorothy P. Rice (*61*). Unfortunately, Rice calculated disease costs in quite aggregate terms, and so the category "diseases of the respiratory system" must be broken down. It seem reasonable to assume that both direct and indirect costs would be proportional to the period of hospitalization (total patient-days in hospitals) by disease category (*62*).

Rice defines a category of direct disease costs as including expenditures for hospital and nursing home care and for services of physicians, dentists, and members of other health professions. "Other direct costs" (which would add about 50 percent to those just enumerated) consist of a variety of personal and nonpersonal expenditures (such as drugs, eyeglasses, and appliances), school health services, industrial in-plant health services, medical activities in federal units other than hospitals, medical research, construction of medical facilities, government public health activities, administrative expenditures of voluntary health agencies, and the net cost of insurance. Since Rice does not allocate "other direct costs" among diseases, we omit it from our cost estimates. However, we conjecture that respiratory diseases represent a substantial portion of this category. Thus, our direct cost estimate is likely to be a substantial underestimate of "true" direct costs (probably more than 50 percent too low).

Estimating indirect cost is an attempt to measure the losses to the nation's economy caused by illness, disability, and premature death. We would argue that such a calculation gives a lower bound for the amount people would be willing to pay to lower the morbidity and mortality rates. These costs are calculated in terms of the earnings forgone by those who are sick, disabled, or prematurely dead (*63*).

The Health Cost of Air Pollution

The studies cited earlier in this article show a close association between air pollution and ill health. The evidence is extremely good for some diseases (such as bronchitis and lung cancer) and only suggestive for others (such as cardiovascular disease and nonrespiratory-tract cancers). Not all factors have been taken into account, but we argue that an unbiased observer would have to concede the association. More effort can and should be spent on refining the estimates. However, the point of this exercise is to estimate the health cost of air pollution. We believe that the evidence is sufficiently complete to allow us to infer, roughly, the quantitative associations. We do so with caution, and proceed to translate the effects into dollars. We have attempted to choose our point estimates from the conservative end of the range.

We interpret the studies cited as indicating that mortality from bronchitis would be reduced by about 50 percent if air pollution were lowered to levels currently prevailing in urban areas with relatively clean air. We therefore make the assumption that there would be a 25 to 50 percent reduction in morbidity and mortality due to bronchitis if air pollution in the major urban areas were abated by about 50 percent. Since the cost of bronchitis (in terms of forgone income and current medical expenditures) is $930 million per year, we conclude that from $250 million to $500 million per year would be saved by a 50 percent abatement of air pollution in the major urban areas.

Approximately 25 percent of mortality from lung cancer can be saved by a 50 percent reduction in air pollution, according to the studies cited above. This amounts to an annual cost of about $33 million.

The studies document a strong relationship between all respiratory disease and air pollution. It seems likely that 25 percent of all morbidity and mortality due to respiratory disease could be saved by a 50 percent abatement in air pollution levels. Since the annual cost of respiratory disease is $4887 million, the amount saved by a 50 percent reduction in air pollution in major urban areas would be $1222 million.

There is evidence that over 20 percent of cardiovascular morbidity and about 20 percent of cardiovascular mortality could be saved if air pollution were reduced by 50 percent. We have chosen to put this saving at only 10 percent—that is, $468 million per year.

Finally, there is a good deal of evidence connecting all mortality from cancer with air pollution. It is difficult to arrive at a single figure, but we have estimated that 15 percent of the cost of cancer would be saved by a 50 percent reduction in air pollution—a total of $390 million per year.

Not all of these cost estimates are equally certain. The connection between bronchitis or lung cancer and air pollution is much better documented than the connection between all cancers or all cardiovascular disease and air pollution. The reader may aggregate the costs as he chooses. We estimate the total annual cost that would be saved by a 50 percent reduction in air-pollution levels in major urban areas, in terms of decreased morbidity and mortality, to be $2080 million. A more relevant indication of the cost would be the estimate that 4.5 percent of all economic costs associated with morbidity and mortality would be saved by a 50 percent reduction in air

pollution in major urban areas (64). This percentage estimate is a robust figure; it is not sensitive to the exact figures chosen for calculating the economic cost of ill health.

A final point is that these dollar figures are surely underestimates of the relevant costs. The relevant measure is what people would be willing to pay to reduce morbidity and mortality (for example, to reduce lung cancer by 25 percent). It seems evident that the value used for forgone earnings is a gross underestimate of the actual amount. An additional argument is that many health effects have not been considered in arriving at these costs. For example, relatively low levels of carbon monoxide can affect the central nervous system sufficiently to reduce work efficiency and increase the accident rate (65). Psychological and esthetic effects are likely to be important, and additional costs associated with the effect of air pollution on vegetation, cleanliness, and the deterioration of materials have not been included in these estimates (66).

References and Notes

1. For a general discussion of inherent problems in handling residuals, see R. U. Ayres and A. V. Kneese, *Amer. Econ. Rev.* 59, 282 (1969).
2. For summaries of studies relating air pollution to health, see J. R. Goldsmith, in *Air Pollution*, vol. 1: *Air Pollution and Its Effects*, A. Stern, Ed. (Academic Press, New York, 1968), p. 547; E. C. Hammond, paper presented at the 60th annual meeting of the Air Pollution Control Association, 1967; H. Heimann, *Arch. Environ. Health* 14, 488 (1967). For more general reviews of the literature, see A. G. Cooper, "Sulfur Oxides and other Sulfur Compounds," *U.S. Public Health Serv. Publ. No. 1093* (1965); ———, "Carbon Monoxide," *U.S. Public Health Serv. Publ. No. 1503* (1966); "The Oxides of Nitrogen in Air Pollution," *Calif. Dep. Public Health Publ.* (1966); "Air Quality Criteria for Sulfur Oxides," *U.S. Public Health Serv. Publ. No. 1619* (1967); *Effects of Chronic Exposure to Low Levels of Carbon Monoxide on Human Health, Behavior, and Performance* (National Academy of Sciences and National Academy of Engineering, Washington, D.C., 1969).
3. *Public Health* (Johannesburg) 63, 30 (1963); D. M. Johnson, *Good Housekeeping* 1961, 49 (June 1961).
4. See A. V. Kneese, in *Social Sciences and the Environment; Conference on the Present and Potential Contribution of the Social Sciences to Research and Policy Formulation in the Quality of the Physical Environment*, M. E. Garnsey and J. R. Hibbs, Eds. (Univ. of Colorado Press, Boulder, 1967), p. 165; R. G.

Ridker, *Economic Costs of Air Pollution* (Praeger, New York, 1967); "Air Quality Criteria for Sulfur Oxides, *U.S. Public Health Serv. Publ. No. 1619* (1967), pp. 54–57.

5. L. Greenburg, M. B. Jacobs, B. M. Drolette, F. Field, M. M. Braverman, *Public Health Rep.* 77, 7 (1962); M. McCarroll and W. Bradley, *Amer. J. Public Health Nat. Health* 56, 1933 (1966); J. Firket, *Trans. Faraday Soc.* 32, 1192 (1936); H. H. Schrenk, H. Heimann, G. D. Clayton, W. M. Gafafer, H. Wexler, "Air Pollution in Donora, Pennsylvania," *Public Health Bull. No. 306* (1949).

6. See J. R. Goldsmith, *Med. Thoracalis* 22, 1 (1965).

7. B. G. Ferris, Jr., and J. L. Whittenberger, *N. Engl. J. Med.* 275, 1413 (1966).

8. For a summary of laboratory experiments see "Air Quality Criteria for Sulfur Oxides," *U.S. Public Health Serv. Publ. No. 1619* (1967), pp. 79–93.

9. Chronic effects, where the incidence of the disease is small, can be studied only for large samples (millions of man-years of exposure); see J. R. Goldsmith, *Arch. Environ. Health* 18, 516 (1969); J. Rumford, *Amer. J. Public Health* 51, 165 (1961). Morbidity data would be more useful than mortality data, since death may result from a cause having no direct relationship to the original pollution-induced disease.

10. For example, M. McCarroll and W. Bradley [*Amer. J. Public Health Nat. Health* 56, 1933 (1966)] correlate the daily mortality rate in New York City with daily pollution indices. See also J. R. McCarroll, E. J. Cassell, W. A. B. Ingram, D. Wolter, "Distribution of families in the Cornell air pollution study" and "Health profiles vs. environmental pollutants," papers presented at the 92nd annual meeting of the American Public Health Association, New York, 1964; ———, *Arch. Environ. Health* 10, 357 (1965); W. Ingram, J. R. McCarroll, E. J. Cassell, D. Wolter, *ibid.*, p. 364; E. J. Cassell, J. R. McCarroll, W. Ingram, D. Wolter, *ibid.*, p. 367. Other workers have attempted to explain daily variations in hospital admissions [see L. Greenburg, F. Field, J. I. Reed, C. L. Erhardt, *J. Amer. Med. Ass.* 182, 161 (1962); W. W. Holland, C. C. Spicer, J. M. G. Wilson, *Lancet* 1961-II, 338 (1961); G. F. Abercrombie, *ibid.* 1953-I, 234 (1953); A. E. Martin, *Mon. Bull. Min. Publ. Health Lab. Serv. Directed Med. Res. Counc.* 20, 42 (1961); R. Lewis, M. M. Gilkeson, Jr., R. O. McCaldin, *Public Health Rep.* 77, 947 (1962); T. D. Sterling, S. V. Pollack, D. A. Schumsky, I. Degroot, *Arch. Environ. Health* 13, 158 (1966); T. D. Sterling, S. V. Pollack, J. Weinkam, *ibid.* 18, 462 (1969)]; absence rates (see J. Ipsen, F. E. Ingenito, M. Deane, *ibid.*, p. 462); symptoms in school children [see B. Paccagnella, R. Pavanello, R. Pesarin, *ibid.*, p. 495; T. Toyama, *ibid.* 8, 153 (1964); the incidence of asthma attacks [see L. D. Zeidberg, R. A. Prindle, E. Landau, *Amer. Rev. Resp. Dis.* 84, 489 (1961); C. E. Schoettlin and E. Landau, *Public Health Rep.* 76, 545 (1961); R. Lewis, *J. La. State Med. Soc.* 115, 300 (1963)]; and other morbidity [see J. T. Boyd, *Brit. J. Prev. Soc. Med.* 14, 123 (1960); R. G. Loudon and J. F. Kilpatrick, *Arch. Environ. Health* 18, 641 (1969)].

11. The most complete investigation of various pollutants was that of the Nashville studies. See L. D. Zeidberg, R. A. Prindle, E. Landau, *Amer. Rev. Resp. Dis.* 84, 489 (1961); L. D. Zeidberg and R. A. Prindle, *Amer. J. Public Health* 53, 185 (1963); L. D. Zeidberg, R. A. Prindle, E. Landau, *ibid.* 54, 85 (1964); L.

D. Zeidberg, R. J. M. Horton, E. Landau, *Arch. Environ. Health* 15, 214 (1967); ———, *ibid.*, p. 225; R. M. Hagstrom, H. A. Sprague, E. Landau, *ibid.*, p. 237; H. A. Sprague and R. Hagstrom, *ibid.* 18, 503 (1969). It is conceptually possible to differentiate among pollutants, since, for example, the correlation between mean level of suspended particulates and mean level of sulfates for 114 U.S. Standard Metropolitan Statistical Areas is only .20.

12. L. B. Lave, "Air pollution damage" in *Research on Environmental Quality*, A. Kneese, Ed. (Johns Hopkins Press, Baltimore, in press).

13. D. J. B. Ashley, *Brit. J. Cancer* 21, 243 (1967); C. Daly, *Brit. J. Prev. Soc. Med.* 13, 14 (1959); J. Pemberton and C. Goldberg, *Brit. Med. J.* 2, 567 (1954); P. Stocks, *ibid.* 1, 74 (1959); R. E. Waller and P. J. Lawther, *ibid.* 2, 1356 (1955); ———, *ibid.* 4, 1473 (1957); P. J. Lawther, *Proc. Roy. Soc. Med.* 51, 262 (1958); ———, *Nat. Acad. Sci. Nat. Res. Counc. Publ. No. 652* (1959), pp. 88–96; ———, *Instrum. Pract.* 11, 611 (1957); J. Pemberton, *J. Hyg. Epidemiol. Microbiol. Immunol. (Prague)* 5, 189 (1961); J. L. Burn and J. Pemberton, *Int. J. Air Water Pollut.* 7, 5 (1963); E. Gorham, *Lancet* 1958-I, 691 (1958); P. Stocks, *Brit. J. Cancer* 14, 397 (1960). These studies are updated and summarized in S. F. Buck and D. A. Brown, *Tobacco Res. Counc. Res. Paper No. 7* (1964).

14. W. Winkelstein, Jr., S. Kantor, E. W. Davis, C. S. Maneri, W. E. Mosher, *Arch. Environ. Health* 14, 162 (1967).

15. International Joint Commission U.S. and Canada, "Report on the pollution of the atmosphere in the Detroit River Area" (Washington and Ottawa, 1960).

16. T. Toyama, *Arch. Environ. Health* 8, 153 (1964).

17. F. L. Petrilli, G. Agnese, S. Kanitz, *ibid.* 12, 733 (1966); A. Bell, in *Air Pollution by Metallurgical Industries*, A. Bell and J. L. Sullivan, Eds. (Department of Public Health, Sydney, Australia, 1962), pp. 2:1–2:144.

18. P. Stocks, *Brit. Med. J.* 1, 74 (1959).

19. ———, *Brit. J. Cancer* 14, 397 (1960).

20. D. J. B. Ashley, *ibid.* 21, 243 (1967).

21. That the least-squares method provides the best linear unbiased estimates is the conclusion of the Gauss-Markov theorem, for which $E(ee') = \sigma^2 I$ and $E(e) = 0$ are the basic assumptions. These assumptions are that the basic model must be linear and that the distribution of the errors must have an expected value of zero, have finite variance, have a constant distribution over the various observations, and be independent. In addition, no explanatory variables may be omitted which are correlated with included variables. It is also convenient to assume that the explanatory variables are measured without error, although the framework can easily be adusted to handle errors. In order to perform significance tests, one must make an assumption about the distribution of the error term. For all the relations we estimated, we plotted the residuals and discovered that all distributions were unimodel, symmetric, and basically consistent with the normal distribution. Thus, in the discussion that follows, we have assumed that the error term is distributed normally.

22. For example, for economic level 1 (defined below), the death rates (per 100,000) for pollution levels 2 to 4 (defined below) are 126, 271, and 392. For economic level 2, the death rates for air pollution levels 1 to 4 are 136,

154, 172, and 199. For economic level 4, the death rates for pollution levels 1 to 3 are 70, 80, and 177. The five economic levels, based on median family income in a census tract, are as follows: $3005–$5007; $5175–$6004; $6013–$6614; $6618–$7347; and $7431–$11,792. The four air pollution levels (in micrograms) of suspended particulates (per cubic meter per 24 hours) are as follows: less than 80, 80–100, 100–135, and more than 135.

23. P. Stocks and J. M. Campbell, *Brit. Med. J.* **2**, 923 (1955).

24. In most of the early studies, pollution measures were not available, and so urban mortality rates were contrasted with rural rates. In these studies a substantial "urban factor" was found, which, unfortunately, was a compound of air pollution and many other factors. In the later studies the portion ascribable to air pollution is separated out.

25. C. Daly, *Brit. J. Prev. Soc. Med.* **13**, 14 (1959).

26. Buck and Brown [*Tobacco Res. Counc. Res. Paper No. 7* (1964)], in examining data from England, control for population per acre, for social class, and for smoking habits. They find no relationship between smoking and lung cancer, and a relationship between SO_2 and lung cancer that is not consistent. Stocks uses three sets of data to isolate the effect of air pollution on lung cancer. Contrasting data for eight northern European cities, he finds a correlation between lung cancer and air pollution of .60, and correlations between lung cancer and smoking that range between .27 and .36. Contrasting data for 19 countries, he finds that an index of solid fuel consumption is a much stronger variable than cigarette consumption per capita. Finally, with data from northern England, he finds confirmation of an association between lung cancer and air pollution. See P. Stocks, *Brit. J. Prev. Soc. Med.* **21**, 181 (1966).

27. E. C. Hammond and D. Horn, *J. Amer. Med. Ass.* **166**, 1294 (1958).

28. W. Haenszel, D. B. Loveland, M. G. Sirken, *J. Nat. Cancer Inst.* **28**, 947 (1962).

29. Haenszel and Taeuber analyzed data for 683 white American females who died of lung cancer, and for a control group. They found the crude rate of death from lung cancer to be 1.32 times as high in urban areas as in rural areas for 1958–1959 and 1.29 times as high for 1948–1949 (in subjects 35 years and older, with adjustments made for age). When adjustments were made for both age and smoking history, the ratio was 1.27. This ratio increased with the duration of residence in the urban or rural area, from 0.80 for residence of less than 1 year to 1.76 for lifetime residence. See W. Haenszel and K. E. Taeuber, *J. Nat. Cancer Inst.* **32**, 803 (1964).

30. L. D. Zeidberg, R. J. Horton, and E. Landau [*Arch. Environ. Health* **15**, 214 (1967)] are not able to isolate an air pollution effect on mortality from lung cancer from data for Nashville for the years 1949 through 1960; C. A. Mills [*Amer. J. Med. Sci.* **239**, 316 (1960)] investigated rates of death from lung cancer in Ohio. Stratifying according to the amount of driving done by the decreased, he found that the death rate varied with driving and urban exposure; L. Greenburg, F. Field, J. I. Reed, M. Glasser [*Arch. Environ. Health* **15**, 356 (1967)] investigated 1190 cancer deaths that occurred on Staten Island between 1959 and 1961 and found a relationship between lung cancer and air pollution; M. L. Levin, W. Haenszel, B. E. Carroll, P. R. Gerhardt, V. H. Handy, S. C.

Ingraham II [*J. Nat. Cancer Inst.* **24**, 1243 (1960)] found significant differences between urban and rural mortality rates (for periods around 1950) in New York State, Connecticut, and Iowa. For males, the death rates were 41 percent higher in urban areas in New York, 57 percent higher in Connecticut, and 184 percent higher in Iowa. For females, the differences were 7 percent, 24 percent, and 47 percent, respectively; P. Buell, J. E. Dunn, L. Breslow [*Cancer* **20**, 2139 (1967)] utilized 69,868 questionnaires covering 336,571 man-years, in their study of lung cancer in California veterans. They found rates of death from lung cancer (adjusted for differences in age and smoking habits) to be 25 percent higher in the major metropolitan than in the less urbanized areas. Among nonsmokers, the rates of death from lung cancer were 2.8 to 4.4 times as high for major metropolitan areas as for more rural areas.

31. P. Buell and J. E. Dunn, Jr., *Arch. Environ. Health* **15**, 291 (1967).

32. W. Winkelstein, Jr., and S. Kantor, *ibid.* **18**, 544 (1969).

33. For economic level 2 (see 22), the mortality rate per 100,000 for gastric cancer in white males 50 to 69 years old changed from 45 to 41, 48, and 84 as the pollution level (see 22) rose. For economic level 4, the rates were 15, 38, and 63 for the first three pollution levels. For white women 50 to 69 years old, the death rates for economic level 2 were 8, 18, 25, and 40 per 100,000. For economic level 4, the death rates were 5 and 21 for the first two pollution levels.

34. R. M. Hagstrom, H. A. Sprague, E. Landau, *Arch. Environ. Health* **15**, 237 (1967).

35. The four measures of pollution are suspended particulates (soiling), dustfall, SO_2 and SO_3. For all cancer deaths, the number per 100,000 for middle class residents (defined to include about 75 percent of all residents) fell from 153 for high-pollution areas, to 130 for moderate-pollution areas, to 124 for low-pollution areas when a soiling index (concentration of haze and smoke per 1000 linear feet) was used to classify air pollution. When SO_3 (milligrams per 100 square centimeters per day) was used as a basis for classification, the corresponding death rates were 150, 129, and 145, respectively. With dustfall as a measure, the figures were 145, 130, and 131, and with 24-hour SO_2, in parts per million, they were 141, 129, and 138.

36. M. L. Levin, W. Haenszel, B. E. Carroll, P. R. Gerhardt, V. H. Handy, S. C. Ingraham II, *J. Nat. Cancer Inst.* **24**, 1243 (1960).

37. P. E. Enterline, A. E. Rikli, H. I. Sauer, M. Hyman, *Public Health Rep.* **75**, 759 (1960).

38. L. D. Zeidberg, R. J. M. Horton, E. Landau, *Arch. Environ. Health* **15**, 225 (1967).

39. When air-pollution level was measured on the basis of sulfation (SO_3, in milligrams per 100 square centimeters per day), the morbidity rates (for white, middle-class males aged 55 and older) were 64.0 man-years per 1000 man-years for high-pollution areas, 34.1 for moderate-pollution areas, and 36.8 for low-pollution areas. Measurement of air pollution on the basis of 24-hour concentrations of SO_2 gave morbidity rates of 47.2, 36.8, and 22.2, respectively. For these same white, middle-class males, in areas of high atmospheric concentrations of SO_3, the mortality rate was 425.6 per 100,000 population; in moderate-concentration areas, 327.41; and in low-concentration areas, 361.97. With SO_2 concentrations as a measure, the corresponding figures were 424.87, 319.19, and 364.93. When soiling (smoke or suspended particles) was used as the pollution index, the figures were 376.65,

339.13, and 399.88, respectively.

40. G. Friedman, *J. Chronic Dis.* **20**, 769 (1967).

41. The effect of air pollution on pneumonia, tuberculosis, and asthma has also been investigated. C. Daly (see *25*) reports simple correlations of .60 for pneumonia mortality and pollution from domestic fuel and .52 for pneumonia mortality and pollution from industrial fuel. For tuberculosis mortality the correlations are .59 and .22, respectively. The death rates for pneumonia rise from 30 to 52 per 100,000, and those for tuberculosis rise from 47 to 89, as one goes from rural settings to conurbations. Stocks (*19*) reports data on pneumonia mortality, by sex, for 26 areas of northern England and Wales. As shown by regressions 27 through 30 in Table 1, there appears to be a strong relationship between a smoke index and pneumonia mortality. The relationship is much stronger for men than for women. C. A. Mills [*Amer. J. Hyg.* **37**, 131 (1943)], in a classic study of wards in Pittsburgh and Cincinnati for 1929–30, reports substantial correlation between pneumonia death rates and local pollution indices. He found the correlation between dustfall and rates for pneumonia mortality in white males to be .47 for Pittsburgh and .79 for Cincinnati. The actual variation in these death rates is 41 to 165 per 100,000 population for Cincinnati and 0 to 7852 for Pittsburgh. Mills argues that omitted socioeconomic variables could not account for these correlations, but he made no attempt to control for such variables in his studies. He also found that death rates fell significantly as the altitude of an individual's residence increased; there was a drop of approximately 10 percent in death rate for every 100 feet (30 meters) of elevation [see also C. A. Mills, *Amer. J. Med. Sci.* **224**, 403 (1952); E. Gorham, *Lancet* **1959-II**, 287 (1959)]. Zeidberg, Prindle, and Landau [*Amer. Rev. Resp. Dis.* **84**, 489 (1961)] studied 49 adult and 35 child asthma patients for a year. They found that the attack rate (attacks per person per day) for adults rose from .070 during days when atmospheric concentrations of sulfates were low to .216 when concentrations were high. In children, the effect of increased concentrations of sulfates was insignificant. Schoettlin and Landau [*Public Health Rep.* **76**, 545 (1961)] investigated 137 asthma patients in Los Angeles during the fall months. They found that 14 percent of the variance in daily attacks (*n* = 3435) could be explained by the maximum atmospheric concentrations of oxidants for that day. These two studies document a strong relationship between asthma and air pollution; Lewis, Gilkeson, and McCaldin [*Public Health Rep.* **77**, 947 (1962)] found no association between the daily frequency of visits to charity hospitals for treatment of asthma attacks and measures of air pollution.

42. J. W. B. Douglas and R. E. Waller, *Brit. J. Prev. Soc. Med.* **20**, 1 (1966).

43. A. S. Fairbairn and D. D. Reid, *ibid.* **12**, 94 (1958).

44. L. D. Zeidberg, R. A. Prindle, E. Landau, *Amer. J. Public Health* **54**, 85 (1964).

45. Morbidity rates associated with a soiling index were 140, 122, and 96, respectively, for high, moderate, and low pollution; corresponding rates associated with an SO_2 index were 177, 117, and 81. For white females, morbidity rates associated with an SO_3 index were 169, 134, and 160; with a soiling index, 158, 139, and 127; and with an SO_2 index, 172, 136, and 116. For nonwhite males, the morbidity rates associated with an SO_3 index were 86 for high concentrations and 84 for moderate

and low concentrations; corresponding rates associated with a soiling index were 94 and 67, and with an SO_2 index, 84 and 88. For nonwhite females, morbidity rates of 136 and 140 were associated with high and with moderate and low SO_3 concentrations, respectively; corresponding rates associated with soiling were 140 and 129, and with SO_2 concentrations, 145 and 126. The effects for working women and for housewives, between the ages of 14 and 65, were similar.

46. E. C. Hammond, paper presented at the 60th annual meeting of the Air Pollution Control Association, 1967.

47. S. Ishikawa, D. H. Bowen, V. Fisher, J. P. Wyatt, *Arch. Environ. Health* **18**, 660 (1969).

48. W. W. Holland and D. D. Reid, *Lancet* **1965-I**, 445 (1965).

49. D. D. Reid, *ibid.* **1958-I**, 1289 (1958).

50. C. J. Cornwall and P. A. B. Raffle, *Brit. J. Ind. Med.* **18**, 24 (1961).

51. F. C. Dohan, *Arch. Environ. Health* **3**, 387 (1961); —— and E. W. Taylor, *Amer. J. Med. Sci.* **240**, 337 (1960).

52. H. A. Sprague and R. Hagstom, *Arch. Environ. Health* **18**, 503 (1969).

53. L. B. Lave and E. P. Seskin, in preparation.

54. "County and City Data Book," *U.S. Dep. Commerce Publ.* (1962); "Analysis of Suspended Particulates, 1957–61," *U.S. Public Health Serv. Publ. No. 978* (1962); "Vital Statistics of the United States (1960)," *U.S. Dep. Health Educ. Welf. Publ.* (1963); "Vital Statistics of the United States (1961)," *U.S. Dep. Health Educ. Welf. Publ.* (1963).

55. For a discussion of the limitations of these studies, see B. G. Ferris, Jr., and J. L. Whittenberger (7) and J. R. Goldsmith, *Arch. Environ. Health* **18**, 516 (1969).

56. See "Smoking and Health Report of the Advisory Committee to the Surgeon General of the Public Health Service," *U.S. Public Health Serv. Publ. No. 1103* (1964), p. 362.

57. This might be explained by noting that farmers tend to be exposed to a high level of pollution in the course of their work (from fertilizers, insecticides, and the exhaust fumes from farm equipment), which causes more deaths from respiratory disease than would be expected from the low level of general air pollution in rural areas.

58. See, for example, T. Toyama (*16*) and F. L. Petrilli, G. Agnese, S. Kanitz, *Arch. Environ. Health* **12**, 733 (1966).

59. L. D. Zeidberg, R. J. M. Horton, E. Landau, *Arch. Environ. Health* **15**, 214 (1967).

60. R. A. Prindle G. W. Wright, R. O. McCaldin, S. C. Marcus, T. C. Lloyd, W. E. Bye, *Amer. J. Public Health* **53**, 200 (1963).

61. D. P. Rice, "Estimating the Cost of Illness," *Public Health Serv. Publ. No. 947-6* (1966).

62. The category "diseases of the respiratory system" encompasses numbers 470 through 527 of the 1962 International Classification of Diseases, Adapted (ICDA). A report of the Commission on Professional and Hospital Activities, entitled *Length of Stay in Short-Term General Hospitals (1963–1964)* (McGraw-Hill, New York, 1966), gives details on the average lengths of stay and number of patients in 319 U.S. general hospitals for 1963 and 1964 by specific ICDA classifications. From these figures we were able to compute the ratio of total hospitalization by specific disease to total hospitalization for all respiratory diseases. Of the 2,410,900 inpatient days for all respiratory diseases, 232,222 were for acute bronchitis and 177,232 were for "bronchitis, chronic and unspecified." Thus, approximately 17 percent of all inpatient days for respiratory diseases were for some form of bronchitis. On the basis of current hos-

pitalization rates, we find the direct cost of diseases of the respiratory system to be $1581 million annually. An estimated 17 percent of this amount is due to bronchitis; thus, the direct cost of bronchitis is about $268.8 million annually.

63. To calculate the indirect cost of bronchitis, we must do more than take 17 percent of the total indirect cost ($3,305,700) of all diseases of the respiratory system. Almost 50 percent of respiratory disease patients are hospitalized for "hypertrophy of tonsils and adenoids" (ICDA 510). Hospitalization is categorized by age of patient in the Commission on Professional and Hospital Activities report, and we note that 80 percent of these "tonsil and adenoid" patients were under 20 years of age. Thus, it seems clear that the "forgone earnings" of these patients is negligible, and so no indirect costs should be allocated to this group. We therefore excluded the hospitalization of "tonsil and adenoid" patients before computing the percentage of hospitalization due to bronchitis. Thus, we estimated that 20 percent of the indirect cost of respiratory disease can be ascribed to bronchitis.

64. There is one bit of evidence that 25 to 50 percent of total morbidity (and therefore mortality) can be associated with air pollution; see L. D. Zeidberg, R. A. Prindle, E. Landau, *Amer. J. Public Health* 54, 85 (1964). If one accepted this evidence as conclusive, it would follow that the annual cost of air pollution, because of health effects, would run between $14 billion and $29 billion.

65. See J. H. Schulte, *Arch. Environ. Health* 7, 524 (1963); A. G. Cooper, "Carbon Mon-

oxide," *U.S. Public Health Serv. Publ. No. 1503* (1966); *Effects of Chronic Exposure to Low Levels of Carbon Monoxide on Human Health, Behavior, and Performance* (National Academy of Sciences and National Academy of Engineering, Washington, D.C., 1969).

66. Another way to estimate the cost of air pollution is to examine the effect of air pollution on property values. See R. J. Anderson, Jr., and T. D. Crocker, "Air Pollution and residential property values," paper presented at a meeting of the Econometric Society, New York, December 1969; H. O. Nourse, *Land Econ.* 43, 181 (1967); R. G. Ridker, *Economic Costs of Air Pollution* (Praeger, New York, 1967); R. G. Ridker and J. A. Henning, *Rev. Econ. Statist.* 49, 246 (1967); R. N. S. Harris, G. S. Tolley, C. Harrell, *ibid.* 50, 241 (1968).

67. P. Buell, J. E. Dunn, Jr., L. Breslow, *Cancer* 20, 2139 (1967).

68. E. C. Hammond and D. Horn, *J. Amer. Med. Ass.* 166, 1294 (1958).

69. P. Stocks, "British Empire Cancer Campaign," supplement to "Cancer in North Wales and Liverpool Region," part 2 (Summerfield and Day, London, 1957).

70. G. Dean, *Brit. Med. J.* 1, 1506 (1966).

71. A. H. Golledge and A. J. Wicken, *Med. Officer* 112, 273 (1964).

72. W. Haenszel, D. B. Loveland, M. G. Sirken, *J. Nat. Cancer Inst.* 28, 947 (1962).

73. The research discussed in this article was supported by a grant from Resources for the Future, Inc. We thank Morton Corn, Allen Kneese, and John Goldsmith for helpful comments. Any opinions and remaining errors are ours.

Biological Effects
of Urban Air Pollution

L. Otis Emik, PhD; Roger L. Plata;
Kirby I. Campbell, DVM; and George L. Clarke, DVM

Animals were raised in rooms ventilated with
ambient air or air passed through absolute particulate and activated charcoal filters. Average
ambient concentration for 2½ years of oxidant was
0.057 ppm and of carbon monoxide was 1.7 ppm.
Pulmonary alkaline phosphatase in exposed rats
and serum glutamic oxaloacetic transaminase in
exposed rabbits were lower. Pneumonitis was
significantly more prevalent in mice from ambient
air. Male, but not female, C57BL mice survived
significantly longer in filtered air. A/J mice in ambient air were heavier and survived longer. Old
guinea pigs during ten-minute exposures had identical pulmonary resistance to filtered air and equal
increases due to 0.5 ppm of sulfur dioxide, but
those from filtered air showed significantly greater
response to 0.5 ppm of ozone. Decreased running
activity of male C57BL mice was highly correlated
with oxidant concentration.

T HE EFFECTS of naturally occurring air
pollution in Los Angeles have been reported
in a series of articles in this journal under
the same general title we are using.[1-4] Additional reports concerning the studies have
appeared in this and other appropriate journals.[5-10]

The Los Angeles and Riverside, Calif, animal colony studies evolved from an original
proposal to utilize the natural sources of
automotive pollution by placing colonies of
animals "on" the freeways. The Los Angeles
study was started first, utilizing four sites
including one true freeway site between the

inbound and outbound lanes of the Hollywood freeway. At the time that these studies were developed, the downwind location of Riverside in reference to Los Angeles was becoming apparent through the mid to late afternoon arrival of smog "fronts." Riverside would provide a location even more remote than Azusa from the major source of the pollution, permitting more time for photochemical reactions and more opportunity for dilution. Thus, the study essentially established a fifth site in the Los Angeles basin. Similar facilities, similar animals, similar protocols, and similar methods were employed. In fact, the pathological examinations and biochemistry studies were subcontracted to the University of Southern California study in order to insure the comparability and reliability of tissue analyses.

Facilities and Methods

A small animal exposure and laboratory facility was constructed to house this specific study. The 30 × 40-foot building contained two identical 10 × 14 × 8-foot rooms opening to the north. Outside air was drawn through a 4 × 8-foot screened, jalousie window into the ambient-air room (exposure chamber) and exhausted from the inner end of the room by a fan in the roof. Air was forced through one bank of ultrafilters and three banks of activated charcoal into and through the filtered-air room. Ventilation was maintained àt 40 volumes per hour unless the temperature was above 85 F (29.4 C) or below 60 F (15.5 C). At these times, the ventilation was reduced to 10 volumes per hour and a heat pump recirculated room air; the jalousie closed automatically. Animal caging, feeding, and management was essentially identical to the Los Angeles operations.[4]

The original complements of animals were placed in each room in March 1964, and consisted of six male and six female New Zealand white rabbits from a local source used for repeated periodic blood samples and studies of growth pattern; 30 male and 30 female rats, cesarian derived, from Charles River, which were killed periodically for biochemistry studies; 90 male guinea pigs, Hartley strain from Fort Detrick, 50 of which were periodically killed for biochemistry studies and 40 of which were used for growth measurements, for pulmonary function testing on a repetitive schedule, and for determination of survival time; 150 male and 150 female mice, C57BL strain from Texas Inbred, which were used for detection of lung tumors and determination of survival time; 150 male and 150 female mice, A-strain from Texas Inbred, which were killed periodically in order to detect lung tumors; and ten male BALB strain mice which were used in voluntary activity wheels.

The original shipment of C57BL mice was found to be heavily infested with sarcoptic mange mites. All were killed and replaced with a new shipment three months later. A second group of 26 male and 26 female Charles River rats were placed in each room in April 1965. A second group of 25 male and 25 female guinea pigs were placed in each room in May 1965; a shipment of 300 male and 280 female A/J mice were placed in the colony in May 1966. A shipment of 100 male C57BL mice were tested for activity and the 20 which were most uniform were placed in the colony in June 1967, with the next best 20 being added in September.

The publications of Wayne and Chambers,[4] Gardner,[2] and Weg and Wayne[7] report details of methods of killing animals and handling tissues for pathological and for biochemical analyses. Swann et al[9] described the equipment and method for measuring total pulmonary flow resistance in the guinea pigs. Any deviations peculiar to Riverside will be mentioned under the specific result under discussion.

The gaseous components of the outside air and of the two colony rooms were monitored very similarly to those in Los Angeles, as reported by Bryan.[6] Carbon monoxide was measured by infrared absorption; hydrocarbons, by a flame ionization detector; ozone coulometrically utilizing neutral potassium iodide; nitrogen oxides, colorimetrically utilizing Saltzman's reagent; and sulfur dioxide by a continuous conductivity metering technique. The peroxacetyl nitrate (PAN) was measured by electron capture detection, utilizing gas chromatographic separation of air samples injected into the system at 15-minute intervals. The samples were injected by hand during regular working hours until the last five months of the colony when an automated system began to operate the apparatus on a 24-hour basis. Relative humidity and barometric pressure at 1 PM and minimum and maximum temperature were recorded daily in each colony room.

Results

Aerometrics.—A summary of the annual average concentrations for the gaseous agents is presented in Table 1. At times, technical problems with performance of the instrument produced nitrogen oxides data of doubtful reliability. The old sulfur dioxide instrument proved too unreliable to justify reporting the data.

After it was determined that the gases in the ambient- and filtered-air rooms followed the same patterns as had been found by Bryan,[6] the filtered-air room was sampled only periodically to confirm that no oxidant was present. A separate instrument monitored oxidant in the ambient-air room. Compared to Los Angeles,[2,4] the outside concentrations of carbon monoxide, hydrocarbons, and nitrogen oxides were nearly always lower. Oxidant at the facility commonly was slightly lower than that reported simultaneously from downtown Riverside and was generally similar to Azusa. Oxidant exhibited diurnal profiles which had a major peak usually between 2 and 4 PM, with a definite secondary peak or shoulder on the curve at 10 to 11 AM. These shoulders apparently indicated the arrival of local pollution, in comparison to the Los Angeles parcels, which arrived some hours later and were greater in magnitude.

The diurnal curves for PAN were generally similar to those of oxidant, but differ in that PAN may persist through the dark hours at high levels after a strong peak occurs. In the fall, higher peaks and greater persistence of PAN were more common and, compared to oxidant, resulted in higher daily averages, the ratio being approximately 1:12 in the summer and 1:7 in the fall of 1967.

The oxidant level in the unfiltered-, or ambient-, air room was always lower than that in the outside air. For the tabulated data, the regression slope for the ambient-air room was 90% of the outside when the two instruments were functioning properly, but the origin was not zero, the outside instrument averaging about 1.3 pphm higher than the other and varying from time to time.

The Riverside air filtration system used three times the depth of charcoal units, but otherwise was similar to the filtration system in Los Angeles, where Bryan[6] concluded that ozone, nitrogen dioxide, and particulate were removed on a nearly quantitative basis. He found essentially no removal of nitric oxide, carbon monoxide, or hydrocarbons. PAN also appears to be nearly completely removed.

The very different nature of the pollution at the Los Angeles stations[4] in comparison with that at the small animal facility in Riverside can readily be demonstrated. The Azusa station had a five-year average concentration of carbon monoxide of 9.1 ppm compared to 1.7 here, but the oxidant was 5.0 pphm in Azusa and 5.7 here. The Azusa station was described as having more than twice the yield of oxidant per unit of exhaust gas (carbon monoxide being the index of exhaust) of any other Los Angeles station in the animal colony study.[4] On the same basis, the yield here was six times that at Azusa.

Biochemistry and Pathology.—As noted previously, analysis of tissues were conducted in Los Angeles, and the Riverside materials have been partially reported. References 2, 4, 7, 8, and 10 should be consulted for more detailed statements.

In summary, biochemical findings indicate that serum alkaline phosphatase was not different between atmospheres, but levels did decrease with age in rabbits, rats, and guinea pigs. Alkaline phosphatase in lungs of rats was decreased in the exposed groups, the effect being more pronounced in the oldest and in the females. Serum glutamic oxaloacetic transaminase values for rabbits in ambient atmospheres were significantly reduced in comparison to those in filtered air.

Table 1.—*Summary of Daily Average Concentrations of Air Pollutants at Riverside Animal Facility*

Pollutant	Units	Annual Averages			
		1965	1966	1967	Total
Ambient air chamber					
Oxidant	pphm	4.0*	3.9	4.0*	4.0
Pollutants measured outside					
Oxidant	pphm	5.8	5.8	5.5	5.7
Hydrocarbon (as carbon)	ppm	2.2	2.5	2.5	2.4
Carbon monoxide	ppm	2.1	1.4	1.5	1.7
Nitrogen dioxide	pphm	1.9	1.9*	1.8	1.9
Nitric oxide	pphm	1.1*	1.3*	1.9*	1.5
Peroxyacetyl nitrate†	ppb	. . .	3.4	4.6	4.2

* Appreciable data missing, value reliable only for general magnitude.
† Hand sampling each 15 minutes during working hours beginning in June 1966, automatic sampling each 15 minutes for all 24 hours beginning in August 1967.

In summarizing the pathological findings in mice, Wayne and Chambers[4] reported those for the five colony locations as follows: The disorders most frequently observed were acute and chronic pneumonitis, amyloidosis, chronic interstitial nephritis, extrapulmonary tumors and lymphomas. Chronic pneumonitis was most prevalent in the C-57 strain, while acute pneumonitis was noted most often in the A/J mice. In each strain, however, both forms of pneumonitis were more prevalent in the animals exposed to ambient air than in the controls. These differences were statistically significant at the 1% level by the chi-square test when data for all three strains were pooled. In the A-strain mice at Riverside, which were thoroughly examined (87.5% of the 600 lungs and 72.5% of the adrenal glands, heart, kidney, liver, and spleen), amyloidosis was the condition most frequently noted, varying from a low of 5% in the kidneys to 26% in the hearts, but in no case significantly more frequent than in control animals, contrary to reports on mice from Los Angeles. Smog-exposed guinea pigs from all locations exhibited appreciably more pulmonary submucosal calcification, but only in Riverside was cardiac calcinosis not more frequent in exposed animals.

Pulmonary Tumors.—The design of the exposures of A, A/J, and C57BL mice was closely patterned after the studies described by Kotin and Falk[11] with ozonized gasoline fumes as the agent. The outcome of studies in the Los Angeles stations has been reported by Gardner et al,[2,10] who also performed the pathological examinations of the Riverside animals. Generally, the descriptions of the tumors are similar for all sites.

The A-strain mice in the study of Kotin and Falk[11] were killed at the earliest ages and performed differently from any in the later studies. The controls achieved a plateau incidence of lung tumors of approximately 25% by 7 months of age, which was exceeded only at the final killing of the animals when they were 15 months of age. Their exposed mice had 37% incidence of lung tumors at 7 months of age and 80% incidence by 15 months of age. The A-strain mice at Riverside, Table 2, showed a small excess of tumors in the ambient-air exposed group, except at 23 months of age. The incidence of tumors increased steadily from 12 to 18 months of age. The A/J mice showed no consistent or significant difference at any time. Tumor frequency increased with age, but started at a much higher incidence (25% vs 6%).

Multiple tumors were found in the lungs of A-strain mice killed at 16 months of age or older, 26% of the ambient-air exposed, and 29% of the filtered-air exposed, tumor-bearing mice having multiple tumors. All of the A/J mice had multiple tumors when killed, 33% of the ambient-air exposed and 29% of the filtered-air exposed, tumor-bearing animals having multiple tumors at 14 months of age. In animals of the same age, Kotin and Falk found that 61% of the exposed but only 25% of the control tumor-bearing animals had multiple tumors. At 20 months of age, in both sexes and both atmospheres at Riverside, over 60% of the A/J mice had adenomas; in the ambient-air exposed males, the total frequency of adenomas was 1.26 per mouse compared to the maximum of 1.59 reported by Kotin and Falk.[11]

Four C57BL mice with single adenomas were found, two females and one male in ambient air, and one male in filtered air. These findings are similar to those in Los Angeles for the first group, suspicious but not significant. The findings for the later and much larger group in Los Angeles were normal.

Survival.—Of the original complement of 88 male guinea pigs in each room, aliquots of six were killed at approximately 90-day intervals for six periods, the remainder providing survival curves to 1,000 days of age. Of the second complement of 25 males for each room, three from each room were killed. The remaining pigs were used for pulmonary function measurements for two years, after which the survivors were placed in a filtered-air room with additional temperature and humidity control; all were dead within 29 months. Guinea pigs in Riverside and in Los Angeles[1] exhibited no significant differences in survival between the two atmospheres and crossed 50% survival at approximately 550 days.

The A-strain mice were killed in seven batches from 12 to 23 months of age, and the survival rates were appropriately adjusted. At 80% and at 50% survival, Table 3, the males lived longer than females. There was no difference between exposures.

The A/J strain mice were killed in five batches from 14 to 21 months of age. The ages at 80% and at 50% survival, Table 3, both showed that the animals exposed to ambient air survived longer than those exposed to filtered air, and that the females survived longer than the males.

The complete survival curves for the C57BL mice are shown in the Figure. The males and the females show opposite effects of environment. The average age at death for the males exposed to ambient air was 643 days compared to 775 for the males exposed to filtered air, while that for the ambient-air exposed females was 791 and for the filtered-air exposed females was 718. The difference

211

| | A-Strain, 1964-1965 | | | | A/J Strain, 1966-1968 | | | |
| | No. of Mice Killed | | % With Adenomas | | No. of Mice Killed | | % With Adenomas | |
Age, mo	A*	F*	A	F	A	F	A	F
12	24	28	8	4				
14	24	29	20	14	58	58	21	29
16	22	23	32	30	58	60	43	40
18	19	24	50	38	51	58	57	65
20	20	16	35	33	37	44	70	62
21	40	39	45	38	8	23	62	48
23	29	22	55	73				

Table 2.—Incidence of Pulmonary Adenomas in A and A/J Strain Mice at Riverside

* Animals in ambient-air room (A) exposed to natural pollution drawn in from outside; others (F) lived in air filtered through ultrafilters and activated charcoal. A-strain mice born in January 1964; A/J strain mice, in April 1966.

between 775 and 791 is the only one which is not highly significant. The differences were quite accentuated at 80% survival, being 430, 660, 725, and 485 days, respectively. The C57BL mice tested by Kotin et al[12] reached 80% in the filtered air at 550 days, but in the oxidant required only 270 days. The first group of C57BL mice initiated in Los Angeles in 1962 was generally similar to Riverside, the respective figures being 484, 526, 623, and 553 days. Much larger num- bers of mice were tested in Los Angeles starting in 1965. The effect on survival of C57BL mice was consistent in all three sets of exposures: longevity of males was reduced and that of females increased by a lifetime of exposure to air containing photochemical re- action products of motor vehicle exhaust.

Growth.—Information on body weights has been obtained in various phases of our study. The two groups of guinea pigs were weighed monthly at the time of the pulmo- nary function tests. The rabbits also were weighed monthly at the time of blood sam- pling. The A/J mice were killed in balanced aliquots and were weighed at the time of death. The last group of mice in activity wheels were weighed at 14-day intervals as they were changed into clean wheels.

The guinea pigs in the first group (1964- 1965) were nearly identical through 1 year

Table 3.—Comparative Ages at Different Survival Levels for Mice in Different Air Environments

Survival Level, %	Males		Females	
	Ambient	Filtered	Ambient	Filtered
A-Strain, Riverside, 1964				
80	550	555	490	505
50	686	673	663	658
A/J Strain, Riverside, 1966				
80	591	569	612	605
50	610	594	653+	620
C57BL Strain, Riverside, 1964				
80	430	660	725	485
50	725	855	825	805
C57BL Strain, Los Angeles, 1962				
80	484	526	623	553
C57BL Strain, Los Angeles, 1965				
80	435	490	484	472

of age, weighing about 1,000 gm at that time. The animals which lived in ambient air averaged 60 gm heavier in the second year. During this year, approximately 60% of the animals were dying in each room. In the second group of guinea pigs (1965-1967), the pigs which lived in ambient air were 20 to 60 gm heavier in the first year; then the weights became nearly the same as those which lived in the filtered-air room, with a return to the slight advantage of the ambient-air exposed group in the last six months. No significant relationship to temperature or to oxidant could be found.

Ten A/J mice were housed in each cage, and 12 cages in groups of three for each sex in each room were marked to be killed at two-month intervals, beginning when the mice were 14 months of age. At all ages, the animals which lived in ambient air were highly significantly heavier, and, at 14 months of age, the males at 28 gm outweighed the females by 5 gm, narrowing to 2.5 gm by 20 months. Animals with adenomas weighed the same as those with none. This consistent advantage of the ambient-air exposure group is diametrically opposed to the marked weight loss observed by Kotin and Falk in A-strain mice exposed to a 2 ppm concentration of oxidant for 1 year.[11]

A shipment of 25 quadruplet sets of C57BL male sibs, born in February, were tested in May and June, and the most uniform ten full sib pairs were split and placed in wheels in the colony rooms at 18 weeks of age. They weighed 25.5 and 25.9 gm, respectively, for filtered-air exposed and ambient-air exposed mice. Eleven weeks later, the second best ten pairs from the original complement were returned to activity. The original mice now weighed 27.7 and 28.4 gm, while the new mice weighed 26.3 and 27.0 gm, respectively, or 1.4 gm less, which was a highly significant difference. The ambient-air exposed animals were not significantly heavier. In the study of Kotin et al,[12] C57BL mice at 30 weeks of age in filtered-air rooms had just reached 25 gm, while those being exposed weighed less than 20 gm.

Pulmonary Resistance.—The two groups of guinea pigs have been described previously. The apparatus used here was nearly identical to that used in Los Angeles and described by Swann et al.[9] The total expiratory flow resistance measurement includes components from the thorax, lung, and airways, but the animal remains intact and returns for repeated measurements.

The routine monthly measurements made on these guinea pigs failed to reveal any remarkable responses of the animals. In animals from 4 months to 1 year of age, resistance slowly increased in both groups, achieving the average level of 0.4 cm H_2O/ml/sec. At times, the averages were as high as 0.51 but would regress toward the lower value. There was no persistent increase in resistance in either group after the 15th month. In neither group was there ever an observed difference in resistance between those living in filtered and those in ambient air which could be significantly related to any recorded measure of air pollution or weather. However, the oxidant in the room never did achieve the levels reported in Los Angeles

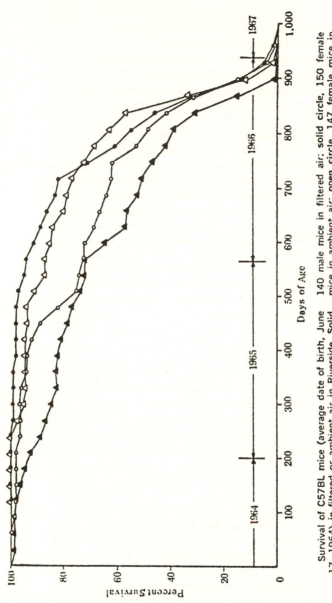

Survival of C57BL mice (average date of birth, June 17, 1964) in filtered or ambient air in Riverside. Solid triangle, 144 male mice in ambient air; open triangle, 140 male mice in filtered air; solid circle, 150 female mice in ambient air; open circle, 147 female mice in filtered air.

and seemed perversely low when measurements were scheduled. Swann and Balchum[3] observed highly significant responses in guinea pigs on two days when oxidant levels reached 0.5 ppm and considered 0.3 ppm to be the threshold level.

At the termination of the study of the second group at 24 months of age, the 12 survivors, seven from filtered air and five from ambient air, were placed for six weeks in a new filtered-air room with controlled temperature and humidity in order to achieve equilibrium. Now pulmonary resistance was identical.

They were then exposed for ten minutes to 0.5-ppm concentrations of the gases PAN, nitrogen dioxide, ozone, and sulfur dioxide at weekly intervals and in that sequence.[13] Increases in resistance of 6%, 17%, 25%, and 62%, respectively, over control were observed, all but the response to PAN being significant. These survivors from the two environments were equally and highly significantly responsive to sulfur dioxide, but those which had lived their lives in filtered air were significantly more responsive to the oxidant gases. This selective adaptation to the oxidant atmosphere among animals with similar responses to sulfur dioxide may be one of the most important observations of the colony study at Riverside. It also may point to one of the major difficulties of epidemiological investigation of health effects. Animals, man, and, perhaps, even plants may become adapted or modified in their immediate responses to exposure, whereby extended and repeated exposure may produce lesser (or greater) responses and complicate the determination of even qualitative, let alone quantitative, associations between exposure and response.

Activity.—Effects of air pollution on voluntary running activity of male mice were included as one of the factors to be evaluated at Riverside but not in Los Angeles. A special stainless steel wheel and cage were developed for this purpose.[14] Ten wheels were used in each colony room.

216

The first set of mice in activity were ten
full sib pairs of BALB males. These mice
were so erratically different from each other
that no useful data resulted. The next set of
20 were selected from 110 C57BL males in
the filtered-air room. Five months after these
mice were placed on test, an accident killed
most of those in the filtered-air room. For
the third set, a group of 24 quadruplet sibs
was ordered and the most uniform ten full
sib pairs were used, based upon a uniform
test of one week. These male C57BL mice
were 113 days of age when placed in the
rooms. Their activity increased for about
three weeks, then leveled off. This group was
followed for the next 13 months and the
findings have been reported.[11] During the
first month, the average daily activity was
23,500 revolutions for mice in filtered air and
21,400 for those in ambient air. By the 13th
month, the averages had fallen to 11,000 and
9,100, respectively. The monthly averages
over the 13 months for filtered activity were
accurately predictable from age and high
temperature $(R = .9959)$. Ambient activity
was similarly predictable from age, high temperature, and peak oxidant $(R = .9958)$.
The difference in activity between the two
rooms was predictable from oxidant alone $(r
= .9533$ for peak oxidant, or $r = .9728$ for
average oxidant). An added correction for
the low temperature difference as well as
peak oxidant gave a slight improvement to
$R = .9651$. When the time intervals for
grouping the data were shortened, the correlation of activity difference with peak oxidant dropped from .953 for monthly grouping to .697 for weekly and to .443 for daily
grouping.

Conclusion

Designed studies were conducted utilizing
populations of laboratory rodents to determine whether the genesis of readily measurable biological effects potentially important to
health would be differently influenced by

217

housing one aliquot of animals in air that had been filtered to remove particulates greater than 0.3μ in diameter, and passed through activity charcoal to almost completely remove ozone, PAN, and nitrogen dioxide, while the other aliquot was housed in ambient air supplied directly from the outside. There was no induction of lung adenomas in A or A/J strain mice exposed to ambient air. Consistent with both Los Angeles groups, C57BL mice did not produce significantly more lung adenomas in the ambient air, and, further, the males in ambient air had significantly shortened survival, but the females had longer survival than those in filtered air. No other significant effects on survival were seen. Growth generally was not significantly influenced by atmospheric environment. However, A/J mice which lived in ambient air were consistently and significantly heavier at the time when they were killed. Guinea pigs which lived in ambient air reacted significantly less to ten-minute exposures to 0.5 ppm concentrations of PAN, ozone, or nitrogen dioxide, but both these and the animals exposed to filtered air were equally and highly significantly reactive to sulfur dioxide. The difference in running activity between two sets of ten full sib male C57BL mice in ambient and filtered air was highly significantly correlated to ambient-air oxidant levels. Generally, under the particular circumstances which prevailed, the joint studies were able to demonstrate a number of biological effects attributable to naturally occurring air pollution in the Los Angeles basin. The absence of differences in some measures and in some of the replications suggests that the exposures generally were near the threshold ability of the system to cause or to perceive effects or both.

This study was supported by Public Health Service contract PH 86-62-28 from the Department of Health, Education, and Welfare. The design and operations were coordinated with a similar study in Los Angeles under the direction of Leslie A. Chambers, PhD, at the University of Southern California. All pathologic examinations and biochemical analyses were conducted there in the laboratories of Murray E. Gardner, MD (pathology) and Myles Maxfield, PhD, (biochemistry). Technicians and veterinarians received instructions and experience at the animal facility under the direction of William M. Blackmore, DVM.

References

1. Swann HE Jr, Brunol D, Wayne LG, et al: Biological effects of urban air pollution: II. Chronic exposure of guinea pigs. *Arch Environ Health* 11:765-769, 1965.

2. Gardner MB: Biological effects of urban air pollution: III. Lung tumors in mice. *Arch Environ Health* 12:305-313, 1966.

3. Swann HE Jr, Balchum OJ: Biological effects of urban air pollution: IV. Effects of acute smog episodes on respiration of guinea pigs. *Arch Environ Health* 12:698-704, 1966.

4. Wayne LG, Chambers LA: Biological effects of urban air pollution: V. A study of effects of Los Angeles atmosphere on laboratory rodents. *Arch Environ Health* 16:871-885, 1968.

5. Bils RF: Ultrastructural alterations of alveolar tissue of mice: I. Due to heavy Los Angeles smog. *Arch Environ Health* 12:689-697, 1966.

6. Bryan RJ: Instrumentation for an ambient air animal exposure project. *J Air Pollut Control Assoc* 13:254-265, 1963.

7. Weg RB, Wayne LG: Automatic analyses of certain enzymes of smog exposed animals. Read before the Sixth Conference on Methods in Air Pollution Studies, Berkeley, Calif, 1964.

8. Maxfield M, Multani S: Effects of Los Angeles smog on mammalian transaminase levels. Read before the 1965 Technicon Symposium on Automation in Analytical Chemistry, New York, 1965.

9. Swann HE Jr, Brunol D, Balchum OJ: Pulmonary resistance measurements of guinea pigs. *Arch Environ Health* 10:24-32, 1965.

10. Gardner MB, Loosli CG, Hanes B, et al: Pulmonary changes in 7,000 mice following prolonged exposure to ambient and filtered Los Angeles air. *Arch Environ Health* 20:310-317, 1970.

11. Kotin P, Falk HL: II. The experimental induction of pulmonary tumors in strain-A mice

after their exposure to an atmosphere of ozonized gasoline. *Cancer* 9:910-917, 1956.

12. Kotin P, Falk HL, McCammon CJ: III. The experimental induction of pulmonary tumors and changes in the respiratory epithelium in C57Bl. mice following their exposure to an atmosphere of ozonized gasoline. *Cancer* 11:473-481, 1958.

13. Emik LO, Plata RL: Comparative effects of four air pollutant gases on respiratory responses of old guinea pigs. *J Air Pollut Control Assoc*, to be published.

14. Emik LO, Plata RL: Depression of running activity in mice by exposure to polluted air. *Arch Environ Health* 18:574-579, 1969.

AUTHOR INDEX